GONE FISHING

GONE FISHING

BOB MORTIMER &
PAUL WHITEHOUSE

Published by Blink Publishing
80–81 Wimpole Street,
Marylebone,
London, W1G 9RE

Hardback – 978-1-7887-019-5-2
Paperback – 978-1-7887-029-4-2
Ebook – 978-1-7887-020-3-4

A CIP catalogue of this book is available from the British Library.

Typeset by seagulls.net
Printed and bound in Great Britain by Clays Ltd, Elcograf S.p.A

7 9 10 8 6

First published by Blink Publishing in 2019.
This edition first published by Blink Publishing in 2020.

Blink Publishing is an imprint of Bonnier Books UK
www.bonnierbooks.co.uk

Bob: To my beautiful wife, Lisa, for all her kindness, fast dancing and wonderful singing

Paul: Dedicated to Harry Whitehouse

CONTENTS

INTRODUCTION

Paul: Fishing. Put simply it's a wonderful way to spend a day or two.

'Simon Peter saith unto them, I go a'fishing.
They say unto him, We also go with thee.
They went forth, and entered into a ship immediately;
And that night they caught nothing.'

The Book of John, Chapter 21: Verse 3

A few years ago, Mortimer and Whitehouse hadn't gone fishing.

Well, that's not quite true. Paul had – he's been doing it all his life; never stopped.

But Paul and Bob had never been fishing together.

And if you're reading this now, you might be like Bob was then.

You might have always wanted to go fishing.

Maybe you thought, 'You know what? I'm thoughtful. I like fish. I don't mind the outdoors. I'm exactly the sort of person who I've always thought would enjoy a good angle.'

You might have dreamed about an imaginary riverbank as you drifted off of an evening, of the dark and tempting shadows gliding gracefully beneath the waters and the weight of these alien shapes tugging on the gossamer thin line that you hold in your hands.

Perhaps you watched *Mortimer & Whitehouse: Gone Fishing* and said wistfully to your companion (human, canine or feline): 'You know what, Alan? That *does* look like fun. Ooh, I'd quite like to give that a go.' And Alan will have rolled his eyes, remembering how you were also really keen on making a cake after you caught the second episode of *The Great British Bake Off*, and how you once told him you were thinking of moving to Bridport while the trailer for *Broadchurch* played in the background.

Maybe you remember fondly messing about in rivers in your youth, which seems both an impossibly long time ago and as vivid as if it were yesterday, and wonder if it would be possible to recapture the wonder of those long-gone days.

You might have been sitting on the Tube, heading into work, and sleepily pictured yourself crunching through an early-morning frost, pushing past a knot of low, twisted trees and finding a suitable spot where the grass has been smoothed down by a previous angler's arse or a slumbering deer. With the noise of coots pottering about and the sound of tiny scurrying from the hedgerows, you can see yourself sitting down, casting out, and before you, the river. Its low and constant bubbling roar, like the sound of blood rushing through your ears, is the soundtrack for your day. Next to you is a Thermos full of hot coffee and a plastic box filled with cling film-wrapped sandwiches, an apple, some Smiths Square Crisps and a Drifter.

And the journey home – wellingtons squeaking as you hoist a duffel bag full of fish onto your back, their damp weight showing the world you've worked hard to put your dinner on the table tonight. And then the train stopped at Euston and you forgot all about it.

You might have pondered buying a rod, or even have one gathering dust in a shed, and you daydream of breaking it out, blowing the dust off and just heading out somewhere – only you don't quite know where yet.

For many people, those idle thoughts and vague hopes are as close as they'll come to fishing.

There are a lot of reasons why so many of us don't ever go fishing. We're busy, we don't know how to get started, we don't know where to go, we don't want to get in trouble …

But it shouldn't be like that.

If you want to, you could go fishing.

If you want to, you *should* go fishing.

But everyone needs someone to give you a helping hand to get you started. An eager friend to help you make that first step.

Hello! Nice to meet you, mate. We're Bob and Paul. And we're taking you fishing!

Fishing has brought us so much enjoyment over the years. It's also given us a TV series, which we made because we wanted to show people why it's such a unique and joyful thing to do.

We were spurred on to doing it because of some health worries, but now we're both as fit as fiddles, we still can't wait to go. What might have started as a bit of physical and mental therapy is now a total passion that we share.

One thing that came out of the series is we'd meet a lot of people who'd watched the show and they would always say the same thing: they'd love to be able to go fishing, but they didn't have a clue how to get started.

So we thought, 'You know what? We should write a book, and then instead of people being sad and unable to fish, we can recommend they pick it up and it will tell them everything they need to know!'

But it's not just the mechanics of putting a hook in the water. A lot of people assume fishing is solely the act of catching a fish. That's a big part of it, but that's not the whole story.

For us, it's also getting out into the beauty of the British countryside: it's spending quality time with an old friend that otherwise you might not see; it's being gifted an inner calm which is harder and harder to find these days; it's the well-earned pint afterwards, one you really feel you deserve, and the dissection of the day just gone; it's the countless delightful little moments that you didn't even realise would appear until you're there, right in the thick of it, and BOOM, there they are; it's just a wonderful way to spend a day. Or two. Or more.

When we decided to write this book, we didn't just want to give you a beginner's manual on how to fish. We wanted to tell you *why* you should fish. And so we're going to tell you why we fish and why it means so much to us both. We wanted to celebrate this ancient pastime that brings its devotees so much pleasure, regardless of where we all are in our respective fishing careers. And hopefully it might resonate with you, whether you're a handy angler or someone who doesn't know a tackle box from a box of chocolates (don't get this wrong on Valentine's Day, by the way).

There's a fisherman inside everyone who's dying to be let out. We hope this book will show you how.

CHAPTER 1
OUR HEARTS

Bob: Can I hold it now, Paul?

'I early learned that from almost any stream…the great secret was this, that whatever bait you used, worm, grasshopper, grub, or fly, there was one thing you must always put upon your hook, namely, your heart. When you bait your hook with your heart, the fish always bite; they will jump clear from the water after it; they will dispute with each other over it; it is a morsel they love above everything else.'

John Burroughs, Locusts and Wild Honey *(1879)*

'But Paul said we did have a reason: *the jeopardy!* One of us might die on the bankside, and if it was filmed, then that would be a feather in the BBC's cap. They'd be able to use it on the news.'

Bob Mortimer, Gone Fishing *(2019)**

* Okay, I know I'm referencing myself and a line from Chapter eight, but it felt very apt.

Paul: It's worth saying upfront: we're very competitive about our heart health issues.

Bob: Well, as always, you trump me, Paul.

P: I do, because I had something really seriously wrong with me a long time before you, Bob.

B: You are the senior patient.

P: That's right. Shall we begin with my near-death experience?

B: Oh, God help us.

PAUL

About 12 years ago, I was on holiday in Somerset with my daughter, just sitting watching the telly or doing something just as wonderful, when BANG!

One minute, I was fine; the next, I thought I'd been shot in the stomach. It was so dramatic and powerful – the sort of indescribable pain where your first and only thought is 'Oh! I'm going to die!' Just brutal.

Apart from knowing this was the worst pain I'd ever been in, I had no idea what was happening. I knew I had to get me and my poor little girl back home, so I left the car there and got on the train with a huge suitcase. It was a Friday night, so the train was packed, of course.

Over the next couple of hours, the pain in my stomach settled down a bit – I mean, it was all relative, it was only slightly less agonising than it had been right at the start – but then I made the huge mistake of eating something. Well, everything you'd imagine would happen to you, if your stomach had basically exploded inside your body and you tried to use it, immediately started up all over again. I'd eaten some crab earlier in the day and even though it had nothing to do with what was happening inside me, I won't eat crab even now. The bad memories are too deeply connected with crabmeat. Never again. If you're a crab, you can bugger right off.

As soon as we pulled into London, I went straight to the hospital. They gave me the once-over, said they thought it was probably appendicitis, but decided to call the MRI operator back into work – she'd gone off for a night out, probably just opened a bottle of wine and about to go off to some party when that phone call came in. She must have been *furious*. Her big night out gets scuppered because someone wants her to come over and have a look inside a screaming man's abdomen.

It was just as well she did, because the scan showed there was something much more seriously wrong with me than appendicitis. It turned out I'd had an abscess that had exploded in my colon, deep in my large intestine, and it had caused all sorts of chaos. You know the phrase 'busting a gut'? That's what it was. I'd literally busted a gut.

They kept me in the hospital that night and I got all the bells and whistles – tubes threading in and out of me, one up me nose, one up me appendage and one into the abscess leading to a catheter. I wasn't about to have any particular procedure – it was an attempt to try and calm me down ahead of draining the bloody thing the next day. But these first hours are the most dangerous ones, as there's a real and constant danger of something suddenly splitting and then everything inside your body instantly ending up in one great pus tsunami. If this happens, what it boils down to is – and I'm going to use a medical term here – you just die.

It's not an over-exaggeration to say there's a high fatality rate with this sort of thing. So despite the horrendous pain, the gallons of vomit and the near-constant internal examinations, it turns out I was actually one of the lucky ones.

In the end – and this may not come as a huge surprise – I didn't die. Instead, the doctors managed to drain the abscess and then put me on hold to see if it would stabilise, because the abscess had also caused all sorts of lesions and problems inside the small intestine. After a few months of hospitalisation just trying to calm it all down, they told me they were going to do a laparoscopy – a keyhole surgery technique where they go in through the abdomen to stitch your insides back together.

If you can't have a laparoscopy, then you get the more drastic alternative: they open your torso up like a human sports bag and perform a colectomy. And even if the colectomy goes perfectly, you'll end up with a colostomy bag for three months until everything inside starts to calm down.

In the event of anything going wrong with the laparoscopy, there's a standby emergency team waiting to wade straight into your belly and then tie you off – in which case, you'll end up having the colectomy and the colostomy bag.

They went as far as marking me up for it, just in case the worst happened: 'Which side would you like the bag, Mr Whitehouse?' Well, it wasn't something I'd given a great deal of thought to over the years, but in the end, I went for the left. Seemed like the obvious choice.

Even when you get to this stage, a lot of people don't survive the rejoining operation, and considering how many of the various steps along the way lead directly to instant death, I was pretty lucky not to die. The doctor even told me a couple of times, 'How you're still here, I don't know,' (which is right up there in the list of things you don't really want to hear a medical professional say to you when you're not well).

I was really in a lot of pain and I realised that the speed at which they rushed me through the hospital was a sign of how dangerous things had become – but at the time, you're being gripped so hard by the sheer brutality of the experience, you don't really comprehend how risky it actually is. Even now, I don't like to look back on this time too much – it's much easier to use my inherent shallowness and block it all out. But I know I was very, very lucky. So far, so funny, eh?

I remember waking up the morning after the operation, the general anaesthetic slowly wearing off, and the sun was shining and I just thought 'Yeeeeeeessss!' But it wasn't a feeling of triumph or joy – it was more an all-powerful sense of pure and utter relief, because the three preceding days had been beyond awful. Admittedly, I was also completely morphined off my nut, so that probably added to the sensation.

That was my first proper brush with mortality. But it didn't stop there (maybe because I knew one day I'd have to beat Bob). It was while I was being monitored in the aftermath of all that when the doctor noticed my blood pressure was consistently high. I was put on blood pressure medication for a couple of years afterwards, but since it didn't really seem to go down, they brought me back in, did an angiogram and told me, 'Oh, you've got one artery that's only got 10 per cent function.' So, more as a precautionary measure than anything else, they decided to put in a stent.

To put in a stent, they push a camera in through a vein, have a look around, take the camera out, and then the stent gets pushed in along the same route. It goes in deflated and once it's inside and in place, they blow it up with a little balloon. That opens up the steel mesh and then they pop the balloon. Bingo. You've now got a stent leading into your heart. Didn't even leave a scar.

I had three stents put in, which is such a routine procedure I wasn't even knocked out: you get sedated rather than given a general anaesthetic. But having a stent put in is, of course, a delicate and extraordinary procedure. An artery can end up being punctured at any time, so they've got the open-heart team on standby with their hands hovering over the old circular saw. You know that team must be secretly hoping things go wrong so they can start that big saw up. One of the perks of the job.

I stayed overnight because they went in through my femoral artery and an artery can't be stitched, so they had to give me a silicon plug. There was actually more concern about the silicon plug coming out than there was about the heart, and I had more discomfort from that plug than anything to do with the heart.

So I was a bit ahead of Bob in that I had my heart done first. But my operation was fairly precautionary.

It wasn't like Bob's.

For him, it was very nearly all over.

P: Harry [Enfield] and I were on tour, and Bob and Jim [Moir, AKA Vic Reeves] were about to go on tour and we heard – I think Charlie Higson told me – that you'd *presented*, hadn't you? You presented, didn't you, Bob? Who knows *what* he presented.

B: It all happened very fast, Paul. It took a day from presenting to my GP before I was standing in front of the consultant to see if he could put in a stent.

P: I'd been in touch with Bob by this time. 'Stents? Oh, you'll be fine. You'll be back on the beer by tomorrow.'

B: When they told me I had to have a serious heart operation, my main memory is standing in my kitchen and thinking what

I was really going to miss was my little tea towel. I also had an egg cup I was going to miss.

P: When you say you'd miss your egg cup and your tea towel, that's a really poignant and interesting observation – but everyone you say it to always goes 'ha ha ha ha!' And actually *no*, you do really mean that – 'I'll miss these tiny little things.' And it's so difficult for you, Bob, to express that without everybody just laughing.

B: It really struck me that not for one minute did I think, 'Oh, I'm going to really miss performing, I'm really going to miss working.' The things you're going to miss are your wife, your egg cup, your seat that you sit in to watch TV…

P: Sorry, your wife? You've never said that before.

B: What would you miss, Paul?

P: *Nothing.* What do you think I am? An old *softie*?

BOB

I actually have a really good heart, sweet as a nut. I have it checked every year, all the valves. My heart is *fit* – it has a beat so regular and intense that it is often sampled by dubstep and grime artists when they need a more natural sound – but I've got heart disease. Paul's the same.

Heart disease is nothing to do with your heart, as such: it's about the pipes that go in and out. It's genetic. And I think I'm right in saying, if you haven't got the heart disease gene, you won't have a problem whatever you do. If you do have it, it seems to store bad cholesterol in your arteries and in the neck.

Back in October 2015, I had a little pain in my chest. Nothing spectacular, no worries as far as I was concerned – my mum would have said, 'You've got a cold on your chest, have some hot Ribena and a Beechams Powder.' But in three weeks Jim and I were about to head off around the country on our 25th anniversary tour of Reeves and Mortimer, so I thought I'd better have it checked out in case I needed any antibiotics or anything.

The doctor listened to my heart, looked a bit anxious, and sent me straight to a heart consultant the next day. The consultant put me on a running machine for an exercise stress test. This involves running as fast as you can for ten minutes with a number of electrodes attached to your chest. If there is any narrowing of the arteries this will show up on an electrical tracing they record during the test.

The consultant looked at the results and said, 'Right, you need something. You're going to need something done.' Well, my heart sank and the qualms invaded my stomach. It was serious – I wasn't just going to be able to pop some pills. Next step would be an angiogram at the hospital. Just like Paul, this is where they insert a camera through an artery in your wrist and have a good rummage around the arteries surrounding your heart. It's very much like when they put a camera down your drains when they are blocked only in my case the smell is potentially worse. His guess was that I would need a couple of stents to open up narrowings in my pipes.

Three days later, I went to Pembury Hospital with Lisa, my partner of 25 years. I wasn't overly worried. I had spoken to Paul, who had reassured me that having stents inserted was a perfectly simple and pain-free procedure. The stents would be put in as part of the angiogram procedure and I would be as right as rain in a week and ready to go on tour. The room where I went to have the stent was amazing – the most efficient, military, humans-working-at-their-full-capacity room I've ever seen. This was the first time I got scared: it suddenly felt very serious.

After about two hours the procedure was over and the camera was pulled out of my wrist. The consultant, Mr Lawson, explained that I had significant heart disease and some of my arteries were 95% blocked. These blockages could not be stented. He said, 'You need a heart bypass. You've got to have open-heart surgery.'

That was the moment. And I *shit myself*.

There is a theory in showbiz circles that mentally you remain the same age as when you first tasted fame, and I think there is a grain of truth in this. I had been living my life like a bloke in his mid-thirties, drinking, smoking, having a daft laugh and messing about. As I lay on my gurney being wheeled back to my room, I suddenly felt very old and very vulnerable. I became pathetic.

Knowing what I know now, it's not terrifying at all in terms of the risks of the operation – but I think I was so terrified because heart operations were such big things, back when I was young. They were these massive operations that a lot of people didn't survive.

Back home, I spent the evening crying and hugging Lisa. My whole world had suddenly shrunk to include only those things that really mattered to me. Just the sight of little everyday things such as my favourite egg cup or a tea towel that Lisa and I had bought years ago would set off the tears. I became hyper-aware of every little beat and twitch of my heart, convincing myself that it was about to blow. Every time I looked at Lisa I thought of the thousands of kindnesses she had done for me over the years. I asked her if she would marry me and she said yes.

The next day I got down to sorting everything out. I booked in for my operation the following Monday. I phoned up Jim and broke the news that we would have to cancel the tour. Mr Lawson reckoned that if it went ahead, I would have dropped on stage at the Southampton Mayflower Theatre, about ten nights into the tour.

I made a will and went to the registry office with Lisa. The registrar told us that we could not get married without giving 21 days' notice. I explained about the operation and the registrar said, 'Well, you can get an emergency licence if you are literally at death's door.'

So I phoned up my consultant and said, 'Hello, I'd like to get married but I need to prove I'm about to die.' He faxed over a letter to them immediately and he sent a copy to me, but he phoned me up to say, 'Please don't open that letter. *Please.* The information I've put in there is purely to get you your wedding.' He must have laid it on thick. I've still never seen what he wrote; I reckon Lisa scurried the letter away. That's fine, I'm not sure it'll do me any good to read it.

The registry office had to send his letter up to London for some big senior registrar to sign off. The next day the registrar telephoned to say that a licence had been granted and she could marry us on Monday morning, the day I was due in hospital.

I remember it was a wonderfully crisp and sunny winter's day. A perfect day for some winter pike fishing. It was just me, Lisa and her best friend Roma and my two sons, Harry and Tom – that's it. I was wearing the cleanest jacket I could find in the cloakroom cupboard and my sons wore nicely clashing checked shirts. Lisa, as always, looked beautiful. I cried throughout the ceremony. It was one of the happiest occasions of my life.

Afterwards we went to a local café for a fry-up. It would be the last pieces of bacon and the last sausage I would have for a long time. Lisa drove me straight to London after the breakfast meats and I checked in for my operation.

The operation itself took about four or five hours. There are five or six arteries going into your heart and they all branch off again and again. If there's a block in one of those branches, they have to bypass it. They don't get rid of anything; they just skirt round the blockage and join up the working bit with another little working branch. It's ring-road technology, basically.

I had three bypasses in total, using arteries harvested from my leg and the right side of my chest. Basically, they open up your

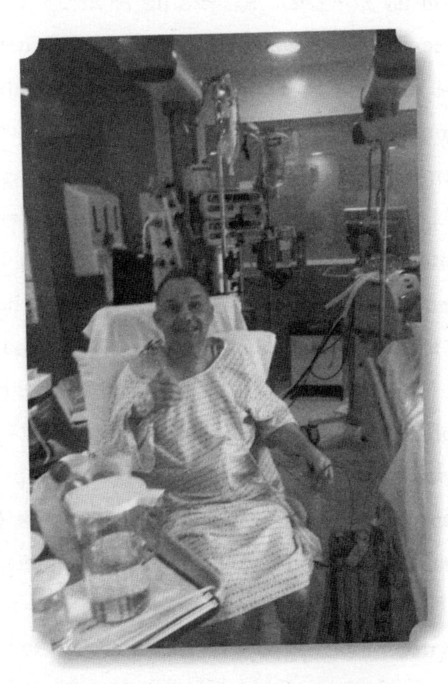

Bob: Here I am during my recovery.

chest with a saw, clamp it open, deflate your lungs, pull out your heart and stop it beating by injecting it with potassium. A machine takes over the functions of your heart and lungs. The bypasses are sewn in and everything put back in place before stapling your chest bone back together. I was in surgery for about four hours and my heart was stopped for 32 minutes. My surgeon, Chris Young, had saved my life. I've thanked him on a number of occasions but it never seems enough.

After the operation I was placed in intensive care for two days. Unfortunately, when they reinflated my lungs they had displaced 40 years' worth of smoker's tar that had built up inside me. I couldn't breathe. There was a TV in my room showing a live penalty shoot-out between my beloved Middlesbrough FC and Manchester United. Just as Grant Leadbitter scored the winning penalty a huge slug of tar trapped in my throat, and the IC nurse shouted at me to 'BREATHE, BREATHE' ...and I was gone.

The next thing I remember is waking up with Lisa by my side. She said, 'Hello, you idiot.' The relief was incredible; I had a second chance. Only ten days after first going to my GP, I was sitting in hospital, recovering from the surgery. An amazing ten days.

So much of your life is unmemorable when you get to my age. Can you remember many days from, say, two years ago? I wouldn't normally remember much six months after: 'Got up, watched telly, went to a meeting, bit more telly.' You might remember, what, five specific days in a year? So, in an odd way, it was quite nice to have a really intense experience.

The second half of the Reeves and Mortimer tour was due to start in 12 weeks. My surgeon and consultant said that I was able to do it if I got exercising and eating properly as soon as possible and, above all, stopped smoking. I'm glad I had this target to aim for as it would have been all too easy to sit on my couch feeling sorry for myself.

I was discharged from hospital after a week. My lungs were still a bit deflated and I would panic at the slightest twitch in my chest or creak from the staples in my sternum. Gradually, Lisa got me up and moving. At first I could only walk the 20 feet to my front gate, but soon enough, we could walk to the park together. Paul would text me every single day, carping on about 'exercise, exercise, exercise'. A month later, and we were slowly jogging together around the lake in the park. And 12 weeks later, I was back doing the tour with Jim.

By then, I basically had a clean bill of health. My lungs were clear, my chest staples had welded to my bone and I no longer craved a cigarette. My cholesterol was down from 8.5 to 2.7 – that was through eating seeds. I'm basically a chaffinch now.

P: It affected you a lot afterwards, didn't it? It did me, big time.

B: It was a *hammer*. Let's say males die at an average of 73 or something – before this happened to me, I'd never thought, 'That's only 15 years away.' It never even crossed my mind. Now, it crosses my mind not far off every day. You know, I think, 'God, if I'm unlucky, that's only *six years*.' You think about it in the funniest of circumstances – like, if I die, I won't see what happens with Brexit, or I'll miss the next World Cup.

P: That was your biggest concern, wasn't it?

B: It really was.

P: I remember you were *terrified* you wouldn't live to see another one.

B: You do think about all these things. I'm forgetting them a bit now, but for a year, I was quite intense. The thought of dying just makes me really sad. I don't feel scared about death, I just feel so frustrated and sad to think I won't see how stories end. My children's story. My wife's. The football. All the stories going on in the world that you're going to miss the end of. It's just so *sad*.

P: So we had to get you out of that state, didn't we, Robert? And what did Dr Whitehouse recommend?

B: While not a qualified doctor, he prescribed the application of fish. And it worked!

*Paul: The pre-*Gone Fishing *days: one of our first trips together following Bob's heart bypass.*

CHAPTER 2
OUR FRIENDSHIP

Bob: For a man approaching 75 years old,
Paul's very youthful isn't he?

Paul: We've only been fishing together in recent years, but Bob and I have known each other for a large part of our working lives. I first saw him when he was doing the *Big Night Out.*

Bob: At the Albany Empire?

P: Yeah.

B: This is *so long ago*. How did we become so *old*?

PAUL

Bob and I have always had a very easy relationship. We've always made each other laugh.

Some people might think we've had quite similar careers, but that's not really the case. After all, I've done films, I'm in the West End, I've done seminal radio shows, and for my TV work, I've won BAFTA after BAFTA …

But despite being much more celebrated by the public and my peers than Bob is – *or ever will be* – I suppose there are some small similarities between us. For example, the way in which both Bob and I became – for want of a better word – *extremely* famous.

Comedy is a strange business. I mean, what on earth compels people to go and stand up in a room full of strangers and say, 'Will you please love me?' Neither Bob nor I had that need. At least, not initially. But thankfully for us, both Harry Enfield and Vic Reeves did.

At the tail end of the 1980s, I was working with Harry, while Bob was working with Jim (Moir, aka Vic Reeves). We were both in comedy partnerships with people who were more than happy to take centre stage, while we were more comfortable being one step back.

It goes without saying that I would never have become a comedian without Harry. I know Harry *would* have been a successful comedian without me – but maybe he wouldn't have been the same comedian he became. I owe so much to Harry and probably don't say it enough. I think you can probably say the same about Jim and Bob.

I'd actually met Jim a few years before he'd even laid eyes on Bob. In the late 1970s, I was living with Charlie Higson in Norwich, studying at UEA – I only lasted a year – and met Jim in passing a couple of times when he came to visit the friend of a friend.

A decade later, I saw him on *The Tube*, hosting a bizarre late-night version of *Celebrity Squares* while he was flying through the air on a wire in a white suit, and I thought, 'Oh, this bloke's in a different league.' I've been a fan ever since.

So at the same time as Jim and Bob were getting together, Charlie and I had started writing for Harry. He was doing Stavros on *Friday Night Live* and it had become really popular so Harry wanted another character. Fortunately, Charlie and I had one up our sleeve: Loadsamoney.

Friday Night Live (and later, *Saturday Live*) gave us the golden opportunity to do that new character on TV for three minutes, and as soon as that first one had gone out – after just one go, just three minutes of it! – Loadsamoney was *everywhere*. Wherever you went, there'd be someone shouting, 'Shut your mouth and look at my wad!'

Writing it was an absolute guilty pleasure. The prevailing mood in comedy at that time was politically right-on, and Loadsamoney's the furthest thing from that you could imagine. I mean, of course he was *wrong*, but there was something gloriously liberating about writing him. Someone once said that Charlie and I spent a lot of time adding fresh depthlessness to that character, and I completely agree with that.

My favourite line we ever did was, 'That Mrs Thatcher, she's done a lot of good for this country.' Of course, the crowd would boo. 'Shut your mouth!' More booing. 'Mind you – I wouldn't shag it.' That single line tells you everything you need to know about Loadsamoney.

Bob always says the best stand-up gig he ever saw was Harry in those early days – just as he was getting famous, when it was three characters in a row, Stavros, Loadsamoney and Buggerallmoney, gag gag gag, bang bang bang! Bob's always been very complimentary about Harry.

When I first went to see the *Big Night Out* at the Albany in about 1989, Charlie and I were quite well established as writers. That night, however, Jim and Bob had us up to do a couple of turns on Novelty Island – and we didn't quite get it right.

We had a routine entitled David Dudd and his Dead Dad. Basically, Charlie built me this prosthetic headpiece, so he could open my brain and start twiddling with bits of it. He starts to pick up radio reception and I start talking a little bit of Dutch and then sing a bit of 'Are You Lonesome Tonight?'. I mean, on the page, it *does* sound fairly funny, but it was too long and a bit self-indulgent.

Not that it mattered: Vic and Bob invited me back on to Novelty Island later on – this time I was the Slitherer, who was a bloke in a bin bag who slithered around on the floor. In fact, perhaps surprisingly, that was the role that got me my actor's Equity card. The people at Equity had told me, 'If you want to join Equity, we need to see contracts of engagement and we need to see you performing in an act.' Equity cards were really hard to get back then, and without one, you weren't allowed to perform on TV. Thankfully, I was now performing the prestigious role of the Slitherer. I don't know whether the Equity bloke even came down to see it or just thought, 'I'm not going to watch the Slitherer, so just chuck him an Equity card,' but either way, they let me join. I was so grateful for that break, I ended up doing the Slitherer on telly when Jim and Bob got their first Channel 4 series.

Some people might look at what Bob and Jim do and think, 'Oh, anyone could do that.' No chance, mate. There's no one else who can do what they do – only them. It's almost impossible to enter their world. He pretends not to be clever, but he's a very clever boy, Mortimer. Much cleverer than he lets on.

I don't have a problem with the Oxbridge lot, but whenever we talk about comedy, Bob always says, 'Paul, have you noticed how people these days want to do their routine and want to be remembered as being clever? What's wrong with just being *daft*?' One of the cleverest things – and Bob won't thank me for saying it's clever – is, 'Oh, Vic … I've fallen.' Who else would think of that? *Nobody*.

He's very much his own man, Bob. He's quite obsessive about work and really into the detail – probably a bit more than me. In my case, I just come up with the ideas and get other people to do it for me: '*The Fast Show* – get on with it. *Down The Line* – start work. *Mortimer & Whitehouse: Gone Fishing* – sort it out, please.'

But that sort of fastidious attention to detail is what I mean about Bob being a lot cleverer than he pretends to be. He's not just the affable fool, he's not just cuddly Bob. I mean, you don't have a 35-year career at the very top of the comedy business *by chance*. You've got to be *so* good, and he is. This applies to me too, but I'm just too modest to say so. And Bob definitely won't.

Bob and I became good mates in Montreal. We were doing the Just For Laughs Festival in 1993, which is this big international comedy showcase. I went with Harry to do 'Smashie and Nicey', and oh, we were so nervous. Well, I was. There was a lot of pressure doing that festival – there were a lot of important industry people from all over the world there, so you walk around the whole time feeling like something's pressing down on your head.

You only do five minutes, but the night before our set Harry and I did this warm-up gig in a little club in Montreal in front of all our mates, and it went down a *storm*. That night, all the English comedians were there and, for once, all the comedians were united in wanting each other to succeed. For the first time, no one was saying, 'Ugh, she's rubbish' – there was none of that. There was a real sense of unity. And so I was really confident! I skipped out, thinking, 'Oh, tomorrow night is going to be *great*!'

Cut to the next night, and we're doing the same routine in front of 3,000 French-Canadians who've never heard of Smashie and Nicey and (surprisingly) have an extremely limited working knowledge of what mid-1990s British DJs were like. And oh, I was *terrible*.

As soon as we started, and the audience went completely quiet, I thought, 'Right, we're in trouble here.' A couple of minutes in and I could feel my mouth drying up. I started to have trouble speaking, and at one point, I nearly swallowed my bloody false teeth trying to say the word 'Fergie'. Harry was great, of course, and got us through it, but I was so glad to get off the stage and get out of there. As I came off, thinking, 'Ugh, that was rubbish,' I remember seeing Bob backstage, who said to me, 'Oh, very good that, Paul,' and I thought, 'No, you don't mean that, you bastard. You're just pleased that you're now going to do better than we did!'

And off he went to do his set – and the exact same thing happened to him! The audience didn't go for it at all, it completely sailed over their heads, and when Jim and Bob came off after their set, Jack Dee said to him, 'Yeah, well done, great stuff,' and Bob thought, 'Oh, you *sh*t*.'

I went back into the audience to watch them because I'm such a fan (I think, sometimes, to the annoyance of Harry, because I used to bang on about them so much). They came back out with a tiny set of The Monkees on a dinner plate and then sang 'Lucky Carpet' in front of these stunned Canadians – 'Thirty foot of beige good luck! Ten square yards of underfelted good fortune!' The Canadian audience had no idea they were watching two geniuses at work, and were confused, unsettled and, in one or two instances, clearly scared. It was *glorious*.

It was like the times I've seen Johnny Vegas go down a blind alley and I've taken real pleasure in watching him do that – basically, it's the delight of seeing a comedian losing the audience, intentionally or otherwise, and thinking, 'Ha ha, now get out of that!' Johnny knows what he's doing and is brilliant at thinking on his feet, but there's something so funny about it. And it was the same that night with Jim and Bob – they always genuinely make me laugh, only this time there was the added piquancy of the whole audience thinking, 'What *is* this?' as they danced around with a big beige carpet. I enjoyed that performance on both levels, but oh, we were all very glad to get out of Montreal.

In 1998, after a decade where we'd both won countless awards for our timeless comedy shows, we did a joint tour of *The Fast Show* and *Shooting Stars*. In those days, people didn't play the O2 because *it didn't exist*. So we did 32 nights at Hammersmith Apollo in a row. I'm still proud of that.

If I'm being honest, when it came to those shows – well, false modesty aside – we *smashed them to pieces*. We blew them out of the water; put them right in their place. It was like a giant treading on a leprechaun. A pike savaging a hapless roach that had blithely swum past it in a pond in the middle of a roundabout next to the Hammersmith Flyover. Even now I keep saying to Bob, 'Why don't we do it again? Me and Harry, you and Jim? Do a half each, it'd be great, what a laugh!' But we don't quite have the, erm, stamina nowadays: you realise, after all, that as you get older, doing an hour is about all you'll be able to manage. But we could do little bits on each other's shows, since it's so great to do it live. I mean, all those 80s bands revive their stuff and go on tour because they think, 'Well, what else am I going to do?' So why not us too? I've mentioned it to the other guys, but whether it'll happen, I don't know. I think it'd be lovely. I'd love to do that. We'll see.

The only concern I would have about doing a show with him nowadays is that Bob does talk about the arse *way* too much now. He never used to – this is a modern thing. I do try and keep him on the straight and narrow with it because I'm shocked at how anus-obsessed Bob has become in his dotage. Entertained, but shocked.

But he's very funny, isn't he? He's *compelled* to be funny. Presumably because of his other shortcomings in life. You know, he's had to be, hasn't he? In fairness, I could say the same about myself – my shortcomings, I mean. Maybe it's the same for all comedians.

So we're old friends now. And he's great, Bob.

Just great.

BOB

For all the years I've known Paul – and it's probably something silly, like 30 years – when we meet up, we meet up to have a laugh. That is our friendship. We've always made each other laugh.

As my great fishing companion and mentor has mentioned, I first met Paul when *Vic Reeves Big Night Out* was taking shape for television. Jim and I were performing the show every week at the Goldsmith's Tavern in New Cross, doing stuff like putting a hat on, holding up bananas and making 'predictions'. I think we were called Bananaro and Avocadaro.

Thankfully, there was so much goodwill from the audience – people were coming for a night out, not for the show. All the like-minded people down in that little bit of London, with the daft haircuts and the dressing up; it was a safe place for them to be. It really didn't matter what was on the stage.

Jim and I would be dropping characters in and out, and if they went down well that week, then they were back in the next week. And then suddenly, we'd had enough 'hits', if you like, that those hits became the core characters: Man with the Stick, Graham Lister, Judge Nutmeg – us singing, dancing, chatting or whatever it was.

I remember Paul being in the audience the night we did the Tanita Tikaram dream sequence, where we put a little kids' paddling pool in the middle of the stage, climbed up a ladder and dropped dog food into it. I'm not saying that was amazing – it was *crap*. But that's the sort of entertainment we were laying on. Those shows were more chaotic than you could ever imagine, but in amongst all of that, people started getting interested.*

The show was slowly getting better too, so much so that we moved it to the Albany Empire Theatre. Jim said one night, 'Oh, I've got a couple of mates coming,' and that's where Paul came down with Charlie.

He asked them if they wanted to do a turn, because anyone could do anything for the show. If someone said, 'Can I come up and do something next week?' then the answer was almost certainly, 'Yes, of course you can.' So Paul and Charlie came down and did something. It was called David Dudd and his Dead Dad: Paul sat on Charlie's lap and, as I remember it, pretended to be a ventriloquist's dummy. I *think* they fried some bacon, too.

I know Paul's said he found it hard to fit into what we were doing, and I think there's some truth to that. Paul and Charlie being Paul and Charlie, they'd have known their words, known

* **Paul:** Just to say that I did end up being Tanita Tikaram in her dream sequence.

their jokes, and they would have had a punchline. Their performance would have been thought through. They're classy. And that always felt a little bit not quite right. I mean that kindly. Going into the bubble that was *Vic Reeves Big Night Out*, it would be very easy to think, 'Oh, it's a comedy night, you've got to be *good*.' And that's not quite it. No, no, no; you're meant to be *rubbish*.

It's a weird thing to try to explain. Jim and I have always refused to wonder how it works, because *we haven't got a clue*. All it is is two blokes really getting on, with their little private jokes – only we've made them public. It's s*omething* like that anyway.

People ask us about our influences and how we decide what we're doing, and we just seem rude because we can't answer. We're not being churlish, it's just we do have a little mental block about it. In the first instance, we go to Jim's house and we chat. We usually start off by just saying, 'Did you watch something last night?' and at some point, someone might say, 'Michael McIntyre, he had a really nice bobble hat,' or something, and we'll say, 'Oh, bobble hats! What about a bloke who has a bobble hat and...?' and we just talk. Little words and images just attract us.

You can see clear echoes of the way I work with Jim in *Mortimer & Whitehouse: Gone Fishing*. There's a definite crossover between mine and Paul's comedy and I love what Paul's done. I had to do a TV show recently where you picked your favourite sketches and I looked back at loads. I typed 'best sketches' into YouTube and I came across so many comedy sketches I remember fondly but that really haven't aged well. But all of Paul's stuff, especially from *The Fast Show*, is still very funny.

I suppose there's a slight timelessness in the work both Paul and I make. We're not really observational comics – although we both do subtle observations of people, we sort of deconstruct characters rather than do the obvious thing.

For example, when Mark Williams comes out and says, 'Today I will be mostly eating taramasalata,' it's partly funny because of that character's isolation, and partly because of his insistence that his life's important, but that's not worthy of examining too closely, it's just *funny*. So we're in a similar area, but the big difference between us and Paul's stuff is Paul's like a comedy mathematician. He's so good.

I love the fact that he takes little bits of culture from *such* a wide spectrum – look at an Armstrong and Miller or a Fry and Laurie, and you can kind of see the parameters of their comedy. But Paul's like a deep ocean, full of very different fish (characters) coming back to spawn (appear) in his freshwater rivers (sketch shows). He'll do everything from a very authentic Cockney pit bull owner all the way through to a very upmarket continental surgeon.

He's also a very good actor. Jim is as well. I know he knows I'm doing it, but I like winding up Paul when we're out on the riverbanks by saying, 'You're a really good actor, what are you fishing for? You're a better actor than that man, and you're a better actor than that bloke. Does it not piss you off?' And he doesn't like it particularly, because I think he knows, in a little alternate last 20 years, he could maybe have been a very successful actor. Maybe he still will be. For a man fast approaching 75 years old, Paul's very youthful, isn't he?

If I'm being honest, perhaps because of the significant and quite startling age difference, I always found Paul a little bit intimidating. For example, I've always been really intimidated by people who went to Cambridge and Oxford – I can't help it, it's their confidence; they're people who couldn't care less. When I go and pitch things to Oxbridge people at the BBC and elsewhere, I always find they're so *articulate.* It reduces me to a gibbering wreck. But Paul's not like me, because he particularly loves the challenge of Oxbridge. If he's with Oxbridge people, he'll say, 'I'm going to *dominate* this!' And he makes himself number one.

Plus, he lights up a room when he comes in. He's very gregarious: 'Hello, everyone!' and makes you feel at ease, whether in a pitch meeting or in a pub – he's very comfortable in any situation. If I went somewhere and Paul was there, I'd always think, 'Great, I'm going to be with Paul for the night.'

But the other people I feel intimidated by – and you'll know the type I mean – are those incredibly intelligent working-class lads. They frighten the *life* out of me and I think: 'You're bloody bright, don't want to take you on.' They pop up from time to time, don't they?, the guys who make me think, 'He's funnier than me *and* more intelligent than me – this is difficult.' Mark Lamarr is a good example of that kind of bloke: so bright and so funny. I was very friendly with Mark, but it was like I was the junior in the relationship. He would make me laugh and I'd chip in a bit, but I would never dream of arguing with him.

I think Paul is one of those guys: like there's so much more depth and intelligence to him. For example, I always thought Paul was incredibly left wing as well, which can sometimes make you a bit on edge and worrying that you might say the wrong thing.

But we've never had an in-each-other's-pockets friendship to discuss those things in the past, and once we had kids, I didn't see Paul so often.

You have a life before and life after your kids, don't you? There's the one before, which is being young, going out, meeting people, having a right laugh, drinking, going to discotheques, driving Formula One cars and posing in the pages of *Mayfair*. And then there's the one after, where you're stuck inside your house 24 hours a day, silently wiping a child's sick off your trousers with a tea towel.

When you have kids, you lose touch with a lot of the mates you used to spend a lot of time with. I stopped seeing Jim of an evening. Didn't see Paul for ages. Basically, for the past 18 years, I stopped doing anything but having the kids. And I should say, I stopped quite happily. But once your kids turn 18 or 19 and leave home, and that part of your life is done, you come back out of the other end and it can be a surprise when you realise you've ducked out of your old life for two decades, and pop back in to find yourself right on the verge of becoming a little old man.

So just before I went in for my heart operation, Paul phoned me up. I can remember being in the kitchen and quite a few people had called, but I'd see their name or their message and I just couldn't bring myself to have a chat. Not for any particular reason, it was just the state of mind I was in at that time. But I did pick up Paul's.

It was nice. I didn't know what this thing was, this heart bypass, and I didn't look up much about it online. *No way!* I looked afterwards. Anyway, Paul told me he'd had a stent and we talked about his dad, who'd had the op years before and it was just very reassuring.

He's a very caring bloke, Paul. You might have slipped into a conversation something like, 'Yeah, my foot's buggered, I've got to have an operation on that foot in a couple of weeks,' and even if he hardly knows you, in three weeks' time Paul will message you and say, 'I hope it went all right with the foot.' Incredible.

There's a tendency I've noticed a lot more these days if people do nice stuff like that, and follow up on it, people look for motives and negative things. There may well be motives about it, but I tend now to just take people at face value. And I know that Paul only does it because he's very nice.

After my operation, Paul got in touch with me again and he started monitoring my progress after I'd come home – he was obsessed, saying, 'Oh, you've got to do this, you've got to do that, you've got to go and exercise.' He even started texting my wife: 'Is he exercising yet?' So he registered. You know – 'Oh God, it's Paul again.'

I don't like to put a time frame on it, but I was slowly getting better. First day up, I walked to the end of the drive. The next day, I walked to the first lamp post beyond the drive. The day after, I got as far as the second lamp post. And I kept that up until I was eventually able to make it as far as the train station. But the whole time I was walking, I was thinking my heart was about to give out.

It's amazing how active your heart is, but if you consciously tune into it for five minutes, you'll convince yourself your heart's completely buggered. It's nothing, but if you properly listen, you'll be, 'One beat …two beats … three beats … *what* … what was *that*?' So when you're having a walk after a heart bypass and listening to your heart, it's really easy to convince yourself, 'Oooh, I'd better stop.'

The person who did my bypass has done a *lot* of them and he said to us there are two responses when you've had the heart done. He said, 'Half the people say, "Well, I've had that done and now I'm sitting here *forever*. I'll eat properly, but I'm doing nothing ever again."' The doctor then said, 'What's better for you is to do what the other half do and get going again.'

Now, if it had been left up to me, I would have been in danger. I *love* telly: I would have sat there for three years. My wife would have tolerated me. You've had a big operation, and I could show the scar and say, in a sad little voice, 'I'm not very well, love…'

You really could hide away from it, and because of the business I'm in, I've always got enough to keep me ticking over. I could have easily tricked myself into feeling like I was still doing stuff: 'Right, so I've got a meeting next week … oh, and I've got to write this the week after …' I could *pretend* I was living.

But Paul wanted me to actually *live*. As soon as I came out of the hospital, he started chivvying me along. He said, 'As soon as you're on your feet, you're going fishing; it'll be really good for you,' because he *knew*. 'You're not going to just sit there; no, you're coming out, you're coming out.' Paul knew that what was important was that he got me out of the house, and he was so right.

So, he teased me out. Eventually, I got up and out of the house. I suppose it was one of those sorts of things where the stars align: I have the heart thing, my kids have gone, I'm not bothered about doing any work, Paul gets in touch …

And that's when we went fishing.

CHAPTER 3

CHILDHOOD FISHING

Paul: My first salmon from the River Teifi.

'As no man is born an artist, so no man is born
an angler.'

Izaak Walton, The Compleat Angler *(1653)*

PAUL

When I was young, I spent a lot of time fishing, and I'm aware when I go fishing now, it's a vain attempt to recapture my childhood.

When I was a kid, everyone went fishing. *Everyone.* Well, not everyone: I'd get together with a load of mates from my street or school, be up early, catch the 5:00am train out to Cheshunt and Broxbourne, all laden down with stuff – I used to make my own sandwiches and coffee – and we'd walk for *miles.* We'd be out of breath the whole way and it'd be barely six in the morning, and we'd have no food left because we'd eaten it all on the way there.

As a child, it's the independence that first appeals to you when you go out like that. You're not just going to a football match on the bus with all your mates from down the road, and then going home again – you're going out into a different world. It's an *adventure.*

You've left your usual urban environment and suddenly, you're out in this unfamiliar rural place, and you're immersed in it. You're not walking through it, you're not looking at it – you're *in* it. You're actually *in* that world and you're a part of it, part of the hierarchy of nature, red in tooth and claw. And maybe for the first time in your young life, for that one thrilling day, you feel like you're utterly free.

Bob used to go when he was a kid as well, so it's a similar thing for him. But he sussed out quite early on that during our trips now you can never *quite* recapture it. You're attempting to catch not only fish but recapture something you'll never quite be able to truly recapture, though even now, when I go down to the river, I still feel that sense of anticipation and excitement that I vividly remember having as a kid.

The very first time I went fishing was with my dad on the River Lea. The Lea has always had a significant history in fishing, because it's one of the places where Izaak Walton used to fish some 400 years ago and was probably in his thoughts when he wrote *The Compleat Angler,* which pretty much set down the modern concept of angling. And it's significant for me personally, because it's where I first fished and fishing's the thing I most associate with my dad.

I was five at the time yet that memory is so vivid. My dad had just got a job at a chemical company called Morson's in 1963, in deepest, darkest Ponders End, when we moved to Enfield from the Rhondda. They were the first British company to manufacture poison gas, but I think they'd probably packed that in by the time he took the job.

We fished in a place called St Margaret's on the Lea – once you get there, you've got the navigation canal and the old River Lea winding in and out of the canal, as it were. He had set up this little rod and float for roach, with a small pinch of Mother's Pride for bait – not even a maggot – cast it out, and said, 'When this float goes under, you tell me.' Well, he turned round to give me the rod and that float *immediately* went under. I shouted, 'It's gone under! It's gone under!'

Straightaway. Immediately, there on the end of the line: a little roach. Oh, and a roach is such a beautiful creature – a sparkling silver, iridescent blue and red little thing.

Paul: Me and my dad enjoy a day's fishing together.

Prepare yourself for a lazy cliché: the roach wasn't the only thing that got hooked that day.

As I got older, I started to do more fly fishing with my dad. I'd done a lot of coarse fishing – that was the main thing where I was brought up in Enfield – but the fishing my dad had done in Wales was much more rural and wild.

Having said that, when he was growing up, the local heavy industry meant the river in the Rhondda Valley ran black with coal dust, so there was nothing in it to fish for. It's ironic that now the heavy industry's disappeared, the river's improved so much, it's sparkling – so they've now got this beautiful river full of fish, but all the work has gone.

We'd go back to the Rhondda Valley for our holidays and we started going out to the Usk or the Wye for a day's trout fishing, because those were the fish he was familiar with from his youth. My dad started as a bait fisherman, until one day he saw this bloke with a fly rod walking down the riverbank, and he thought, 'What on earth is that?' He had absolutely no idea, but once he'd got some kit, he took to fly fishing quickly.

Technically, you wouldn't say he was the best fly fisherman, but his strength was he knew his streams from when he was a young kid, so he was very good at fishing those rivers. Even if his technique wasn't perfect, he could always winkle out a nice trout.

Back in those days, when we were growing up, we'd take all the trout we caught and eat them – we were fishing for the pot. We might put a couple back if we'd caught a lot, but that would be our dinner. Taking them back to my auntie's house in the Rhondda, frying them in a bit of butter, tucking into them as they came out of the pan, jumpers for goalposts – oh, you can't beat it. And when he was young, growing up in the Depression and through the war, it was a real bonus to have fish on the table.

I have absolutely vivid, rock-solid memories of those days. I can picture my dad fishing at a very beautiful place called Sennybridge, high up the River Usk, with me on the bank not quite knowing what I'm doing, and him slightly oblivious to me. I can remember catching my first trout: he'd plonked me on the river with a worm and he went downstream to fish a little run, and I remember pulling out a good-sized wild trout. That's absolutely fixed in my mind. I also remember some beautiful days on the River Wye, where I caught my first grayling and a salmon parr – a juvenile salmon that hadn't yet gone to sea. A creature of staggering beauty, that was returned very quickly.

And I used to make him take me to rivers in England that had more coarse fishing. I remember he'd found the venue, I found this little bit where the stream came in, and we had a lovely day catching small chub, roach and dace – it was magical. I was freezing cold, not wearing the right things, like kids do, and my dad had to effectively rub the life back into me. We went to this part of the River Ouse a lot. I can't imagine it's still like it was and I don't want to find out, because it is fixed in my mind.

I fished with my dad throughout my life. Even up until five years ago, we used to go fishing together. It became something to rely on, that we'd always be able to go out fishing. It's something we were lucky enough to enjoy doing together for the whole of our lives.

One year for his birthday, I got him a day every other week on a tributary of the Test, and that for him was heaven. He couldn't believe he was doing it. I know he was definitely pleased with that. He did well for a boy from the Valleys.

I often wonder if the old Freudian thing is true: do I like fishing for itself, or do I like fishing just to please my dad? I think anyone who does a pursuit that their father does asks themselves that question.

The places I fished as a child still have a really strong draw for me. Every time I go to my mum's, I take a little detour up to Enfield Lock. I go to this bit – it's called Ordnance Road, it used to be a small arms factory – and it's *rubbish*. It's an industrial bit of Brimsdown with a power station on one side and a boatyard on the other. There's a bit called the Pike Pool – but I've never seen a pike there, it's just full of crap.

When I do visit, though, I'm *never* disappointed.

At the time there was a garage near us that also sold fishing gear. It's strange but true, and a testament to how popular fishing was. It was difficult to nick the stuff from there but the people who worked there had no idea about the prices, so you could change the expensive prices for cheap price tags and get a good discount.

In *Coming Up For Air* by George Orwell, he went back to visit a pond of his childhood, where there were big carp – but when he returned as an adult, he was crushingly disappointed. Not me. I don't go mad, expecting to see salmon leaping or anything, because it wasn't like that when I was young. It's not very different from the way it was when I was fishing there as a boy.

That bit of the old river where I caught that first little roach with my dad is now a big estate, so built-up you can barely move between the houses. But if I go up there, I can still see that bit of river.

Even now, whenever I see a little shoal of roach swimming down a stretch of water, I still get excited. It's life-affirming. Later on, my mates and I would go further afield. We'd fearlessly travel on bikes, trains and buses in a way that would make modern kids have breakdowns. We joined the LAA (London Anglers Association), which gave us access to loads of waters. My most formative trips – where I learned about freelining for chub and hemp-fishing for roach, swimfeeding and ledgering for barbel – took place on the old River Lea around Kingsweir. I also joined The Kings Arms fishing club in Cheshunt. I was junior champion in 1973!

Bob: The first time I went fishing was with my dad and we went pike fishing in a lake, but I was so tiny. I only have fleeting memories of him catching one – just a tiny glimpse. My main memories of my dad are him beating me with a leather belt. That's the sort of thing you would remember. You remember that.

Paul: Did you tell me your dad wore lederhosen? Do lederhosen come with a belt?

B: No, Paul, he wasn't permanently in lederhosen. He didn't go to work at Fox's Biscuits wearing lederhosen.

P: It would have been good if he had.

B: 'We should do a Swiss range.' He was adamant about it. 'We could call it … ALPINE.' Put the horn on the front with biscuits coming out of it. 'PRRRRRRPPPPP! Have a biscuit!'

BOB

I have such fond memories of going out fishing as a lad. Bicycle, Woolworths rod, a little bag of size 16 hooks, worms from my garden and a Mothers Pride white sliced strawberry jam sandwich wrapped in tin foil.

We used to go to a place outside Middlesbrough called Great Ayton, to a little pond near there, to catch little roach. You could see them through the clear water and just watch them take your maggots.

Or we'd take the Esk Valley line from Middlesbrough to Whitby and stop at Lealholm or Glaisdale to try and catch a little brown trout. Our success was very limited, as was our technique, but I never remember that mattering too much. What was important was loosening those strings that tied you to your home and your family. I suspect I'm still using some of the lines and routines that were first spoken by me and my mates on the banks of the River Esk.

As we got older, the actual fishing was much less part of the day than the journey, the meeting up, the being away from your parents, being part of a secret gang … finding a mucky book, maybe, hidden in the bushes or what have you. But those memories are just as much about friendship and growing up – being at one with your mates – as they are fishing.

There was always a bit of general uncertainty about the day when you went out fishing as a young man. *Mystery*. I'd think, 'I don't quite know exactly where we would be fishing, I don't know what would happen if I got lost or fell in – I just don't know what's going to happen today.'

You don't know if some bigger boys are going to come around the corner and throw your gear in the river, kick you in the head. It's a bit like that, town fishing, isn't it? You're isolated out there, and there's always that little bit of jeopardy, that you could get kicked in.

There was that fighting technique from the skinhead time where they pulled you down by the hair and kneed you in the face. I mean, as a technique, that's probably been around for ages, but it was definitely around then, and they'd come and practise on you.

They must have been important times in my life, because those early places you fished, they have a resonance with you forever. I used to go to this place called the Roach Pool. It was a place you weren't allowed to fish, but its location was passed down from father to son – 'Oh, there's this secret place, you have to get through a fence, go through these woods …' I mean, I'm sure it wasn't true, but you did have the sense of, 'Oh, we shouldn't be here.'

And even when I was 30 – so I hadn't fished for 15 years – if I went back to Middlesbrough, I'd still drive over from my mum's just to have a look at the Roach Pool. Not fishing, just to have a look.

You think of those great coming-of-age films – like *Stand By Me* and that – and whatever that vibe is, that's what I get when I'm at the Roach Pool. Apart from my family home, it's the only place that gives me that melancholic feeling of childhood.

But it's strange because it really wasn't that big a deal in my life. If I look back, I probably went 15 times or something. There are things I used to do much more often – I used to go to the Cleveland Centre shopping nearly every night – but for whatever reason, this is the only place that has a resonance for me.

So when Paul and I go off fishing, are we actually on a search to somehow connect with a place and time from our youth when we were truly happy? Maybe? Probably!

What I do know is that if I knew it was my last day on earth, I'd go back to that roach pool and see if I could just collapse there and become a part of it. It'd put a smile on my face if I could do it there. Just take the worry out of those last few hours.

I'd actually like to go there now.

So fishing is tied in with that nostalgia, but I've always wanted to do it again. If someone ever said to me, 'Would you like to go fishing?' I would think, 'Yeah, I'd love to go fishing,' but I only know one person who's a fisherman, and that's Paul.

So over the years, I'd see Paul and I'd say, 'We've got to go fishing.' And it only took 15 years for us to finally do it.

P: I went back to Enfield Lock a few years ago and I was just looking at the Pike Pool when a woman came out of the bushes on the other side and immediately started bawling me out: 'What you faacking looking at? What you faacking looking at?' I said, 'I'm looking at the Pike Pool.' 'You're a faacking pervert!' 'Piss off, you silly cow, I'm looking at the Pike Pool!' Turned out she was having a piss and thought I'd been watching her. That bit of the Lea's changed a bit since Izaak Walton's time. Or maybe it happened then as well.

CHAPTER 4
GOING FISHING

Bob: I have no idea what we're talking about in this photo. Probably pies.

Down, and bathe at day-dawn,
Tramp from lake to lake,
Washing brain and heart clean
Every step we take.

Charles Kingsley, Letters and Memories *(1856)*

BOB

I remember, probably as long ago as 15 years, I said, 'Paul, I want to go fishing, so let's meet up in town and you can tell me what rod to get, and I'll buy it.' Paul took me to Hardy's in Pall Mall and I bought an 8ft fly rod and reel. We then went to a park nearby – I don't know what it's called, but there's a bit of grass there – and Paul gave me a fly-fishing lesson in the centre of London.

But we never actually got round to going fishing and I never got to use that rod. Finally though, after my bypass, he teased me out. It was the end of a nearly 20-year evolving story of him trying to get me to fish.

I think Paul knew what was really important was that he got me out of the house. The fishing was a close second, but he knew the key thing was that I did a bit more with the remaining years of my life than stay on my sofa watching *Naked and Afraid* on Discovery Channel (which, I would like to point out, is excellent television).

So he took me down to the Test to this little village called Stockbridge, that he knew was a lovely, life-affirming place ('life-affirming', that's a phrase Paul uses, and I've started using it, too). We arrived in Stockbridge and Paul said to me, 'Now, pretend there are no cars parked on the high street, just look around you – how *beautiful* is this place?'

It was. It's a tiny town full of the sort of houses like you used to see on biscuit tin lids; little tumbledown places with carefully tended window boxes and everything covered in climbing roses. There are pubs and church spires and cobblestones and duck ponds and little bakeries and oak trees – it's exactly what you'd think of if someone says the phrase 'banging traditional English village'. It's like a little corner of Olde England that somehow escaped the 20th century relatively untouched. It was *beautiful.*

And it stirred something in me – it's not easy to explain what it was, but a potent mix of nostalgia, historical pride and delight. Paul gets that same feeling from the rivers themselves, but Stockbridge as a whole really did it for me. It was like I'd walked straight onto the set of a Margaret Rutherford movie from the 1950s – movies that I used to watch in the afternoon as a small boy when I was off sick from school. A cup of hot Ribena in my hand and as warm as toast under a blanket. I found myself thinking, 'This *is* great, isn't it? I like this.' And I suspect Paul knew that was how I'd react: 'Come on, Bob, have a look around. You like this, don't you? Isn't this worth leaving the house for?'*

We checked into this hotel he goes to, and that was also beautiful. Low ceilings, wood panelling, cask ales at the bar, the distant chattering in the kitchens as lunch was being prepared and a winding stone staircase to take me to my room with a small leaded picture window overlooking the river. And finally, Paul had the biggie to knock me sideways. We drove up, walked down a little path and then WHOOOSH – it's the River Test.

It's the poshest river in the country, and it's incredibly gorgeous. Stunning. It's where fly fishing truly evolved, so the town of Stockbridge is full of Americans and Japanese, all dressed like Englishmen, with the tweeds on, all recreating the birth of fly fishing with old cane rods, in the place where it all started. Beautiful.

And the Test was so important to me, because I used to watch fishing programmes when I was a kid – and, as a young lad from Middlesbrough, I was never going to get the chance to fish on the Test. The amount it cost, how could I ever afford to do it? That just wasn't ever going to happen. So being there, getting to fish in it – it just left me gobsmacked.

---

* Hello, everyone. Paul here. I just wanted to add to what Bob's saying here with my own thoughts: I remember at that time Bob suddenly seeing a version of England that was beautiful, traditional and worth celebrating and preserving … more so than I did, actually. My concern is more for the rivers and the habitat, but Bob saw something of 'Olde England' that really got to him, I think.

If I had a worry the first time I went fishing with Paul, it was that I'd ruin his days on the river. I can remember saying to the wife, 'I'm going to the Test with Paul, it's a real treat, but I'm worried I'm just going to spoil it for him. I'm going to take up all his time while he teaches me.' I knew the non-fishing parts of the trip would be fine – Paul and I never run out of things to say, and there's always food we want to eat and drink we'll happily pour down our throats – but I was hyper-aware of not wanting to ruin Paul's fishing trip.

Paul's been a superb teacher over the years, but the interesting thing is I didn't really need to learn much to have a good day out on that first trip on the Test. Even now, I still can't cast a fly, really; I'm a lot better since that first time Paul took me, but I'm still a one out of 20 on the scale. But I didn't need to be a great fisherman to enjoy the day, so I didn't have to bother Paul that much.

I remember Paul said, 'Put the rod in your jacket sleeve,' because it stops you bending your wrist when you cast – so that was enough for me all day. It was all about the little achievements. He would say, 'Oh, that one wasn't bad, Bob,' or if we were fishing for grayling he'd say, 'Try over there, put your float this high and then give it a go,' and that was fine. I'm happy if Paul's happy.

I just kind of know when it's time to let him get on with it: 'Right, Paul, I'm going to go further away for a couple of hours, you don't have to …' It's kind of an unspoken understanding: Paul needs to do his proper fishing. He doesn't need me hanging about, because I disturb the water when I'm throwing my line in, and I'm too noisy generally. So I'm always very aware of that with Paul – that he deserves to have his fishing.

That first trip had a big impact on me. When we'd finished, I realised Paul had played a trick on me. I'm sure it wasn't deliberate, it may have been subconscious, but he'd shown me something spectacular outside the four walls of my house. So he had me 'hooked', as it were. Hooked on life again.

I've improved since that first trip, but I'm nowhere near being a fisherman whatsoever. I'm terrible at fishing. Sometimes I'll catch a fish when Paul doesn't, but that's only because there's so much chance. But I'm a great fisherman in the sense that I love doing it. Not in terms of technique and using all the gear – but I am a good fishing man. I love spending a day fishing.

You should judge the best fisherman not by the number of fish they catch in a day, but by putting them on a machine afterwards and judging how good a time they've had. That's the best fisherman. The one who gets the best bankside score for whether they've had a lovely day or not.

I love fishing with a maggot or a worm, because that's what I used to do, just like in *Huckleberry Finn* – that, to me, is what fishing's about. It's how a young man should fish, and remember, I'm trying to be a young man. It gives me as much pleasure as anyone who's catching huge trout with all the expensive kit. I'm just as anxious to catch a fish as anyone else. And I'm probably more overjoyed than most people if I get one.

In fact, I'm always happy when I'm actually fishing. As soon as I believe I'm fishing properly – when I feel there's a fish there, and someone has told me I'm doing my angling correctly – then I'm away for hours. The glance at the float, the monitoring of the stream and the look at a twig is the nearest I've been to … you know when people meditate, they have a word they focus on and repeat? A word, like 'Morocco' or 'Bowie', and that's all they concentrate on. I think fishing is the nearest I get to that.

The float is so *insistent* that you watch it. It turns my brain off. It's amazing – you cast out, sit there watching that float, just concentrating on it … and before you know it, 30 minutes have passed by and you think, 'I'll have another one.' You might have arrived in the early morning, you've cast a few out, and then suddenly it's midday and you barely noticed the time go by. I think in some way, subconsciously, you've refreshed yourself.

When you go float fishing and you know nothing, like I do, you can convince yourself – and you have nothing to base it on whatsoever – you can say to yourself, 'There's a fish there, I reckon. That bit over there, there's a fish there.' That's all about confidence, because as long as you've taken that step, you can put your line in that spot for the next five hours and be perfectly happy.

There might be bugger all there, but if you *believe*, you'll fish away all day. As soon as you start to doubt it, the pleasure meter goes down. There was one time we went out when I didn't believe there was a fish there – and I couldn't enjoy a moment of it. As soon as I'd convinced myself, 'Oh, there's no stupid fish here,' I felt like it was just a waste of time.

You hear people say, 'Fishing's not really about the fish,' but I'm not finding that. Catching a fish is the thing you're hanging your day on, so it's incredibly important – because that's your main reason for going out.

I think the angling community has contrived a bit to make it a truism that fishing's not just about catching fish. It's their well-meaning attempt to say to non-fishing people, 'Oh no, we're not obsessed with fishing, it's not a big deal, it's just a nice day out.' They fear the ridicule of the non-fishing man. But they *are* obsessed, and they shouldn't be ashamed to say so – I think it makes a big difference to a day's fishing to catch a fish. It completes the story. Otherwise you'd have days when you went out to the riverbank without your rod and just sat on the side, staring out at the water.

That said, I can't particularly distinguish between a day where I've caught and when I've not. Some days have a great narrative because of the fish – because you've tried and you've tried, and then you finally get it right and catch one, and it just takes you to a high. A tiddler can be as important on one day as something massive on another.*

When Paul and I were fishing for tench in the TV series, we had a period of about three hours without even the sniff of a bite. I sensed that, as usual, I was ruining Paul's concentration,

---

* Paul: A blank day here and there is inevitable. Despite what Bob thinks, it is okay to have a blank day; in fact, I would say it's almost necessary to stop you becoming a machine, or so obsessed with catching fish that you'll do almost anything. I've seen this first hand with fishing 'experts' whose reputation is entirely built on them outfishing everyone else. They become fiends who bend rules, break rules and cheat to get a fish. They lose their souls and their enjoyment of fishing. Thanks for listening, please enjoy the rest of Bob's chapter.

so I decided to move further up the bank and have a go with my blue fibreglass rod and a maggot under a float. 'I'll show him,' I thought, knowing that I wouldn't.

But surprise, surprise, within ten minutes I'd landed a pretty little roach and that achievement filled me with pride and gave me the opportunity to declare to Paul that he really needed to up his game. A tiny unexpected fish and one I will never forget.

But then there's the other narrative – one where conversations start, evolve, conclude and are forgotten … where a swan pops over to see you and tells you its troubles … where you collect dry sticks to fire up your Kelly Kettle … where a cow wanders over and tells you her name is Lazy Barbara and 'I'm a really big deal around these parts' … where fleeting memories drift in and out of your mind for hours and where eventually and inevitably, you head off to the pub without having caught a thing.

So fishing's not *just* about catching a fish, but it is a big part of it. It's like when I go to the football. I'm stupidly into football – I'll go in the away end, in the thick of it, and before the match, I pop to the pub where I know the Boro lads will be, and we have our little rituals of when we sing, and if, say, we're playing QPR, we'll all puff ourselves up and march down to the ground like we own London. And I enjoy all of that as much as the actual game of football itself.

So fishing isn't about fish in so much as football supporting isn't about football. It *is*. There's a lot of individual and different things you enjoy about it, but with both you want a result at the end of the day, because that just caps everything off.

So it's not only the fish. For me, a very important thing about the day is the location, the actual spot where we're fishing. I'm very much a fan of the overgrown, closed-in-a-bit, little-ish river, where no one can see you, not even the wildlife, and kingfishers mess about, oblivious to your presence, and a weasel nips across, on his way to the post office without even a nod in your direction, and you're just sort of hidden from the world. That's perfect for me. I think Paul's keener on standing in the middle of great big rivers, but it's different strokes for different people.

I like to get closed in, lost in some reeds – that's what I like, that kind of fishing. Paul always says it's not looking at where you're fishing, it's becoming *part* of it. He says that, but I don't believe Paul's ever been a part of that kind of fishing, because he's never still – he's always walking up and down, flicking around, swishing his line. He's just a hunter, isn't he? If we were fishing in a garden pond, I'd like to be the gnome, while Paul would be the rotovator, doing the lawn.

I've not been to any of the locations where we fish on the show before – they're chosen by Paul and our fishing consultant John Bailey – so it's exciting when we first arrive. When we went to the Derbyshire Wye for the first series, that was lovely: the sun coming through the trees, the water clear and shallow, some over-friendly cows – I was just beaming. Beaming with the beauty of it. I'd happily fish on the Derbyshire Wye for the rest of my life and nowhere else.

On one trip Paul took me fishing for perch and when we arrived at the location, it was just like a municipal reservoir, so I was miserable. I said, 'I don't like it, Paul. This is a crap one.' Without the beauty, there was something vital to my enjoyment that was missing from that trip. The beauty is key for me, because it's a reminder that going fishing is a total escape from my ordinary days. Something that sums up the appeal of fishing is when you're heading off on a trip: 'See you, love. Right, bye bye,' and you push the stuff in the back of the car and slam the door, and it's like your regular life has been put on hold. I like the fact that everything you do from the moment you shut that door is of no consequence to anyone in the world, apart from yourself.

A lot of the appeal is also the anticipation of how enjoyable it's going to be when you get there. A fishing trip starts long before you leave the house. When Paul says, 'I'm going to the Test on the 10th, do you want to come?' – that's the start of a fishing trip for me. You get the diary out, have a check and say, 'Yeah, I can come on the 10th.' And the next step is, 'Well, shall we stay at this place?' – you have a look at it – 'Hell, yeah!' You look at the menu, see what beers they've got, analyse the desserts. And then you start working out the journey down, sorting out the tickets, getting your fishing clothes out of the cupboard – you can stretch it out for weeks.

And then you get there and there's that moment like you have when you go to the football – it's a moment you never lose, walking up the steps at a ground and then you get to the top and there's that moment where you first see that bright green floodlit football pitch lying beneath you. It's a brief and uplifting moment, but it's great. You never lose the thrill of seeing it. And it's the same feeling you get the first time the car you're in slows down and brakes in some quiet green nook of countryside, and someone points and says, 'This is where we're fishing.' Oh, that's a lovely moment.

And then there's a little hiccup, which is setting the stuff up. Bittersweet's not the word, but you just want to get stuck in. Anytime you have to take your shoes off is a bit of a downer. And all the time, the fish will be on the rise, and every second you're not down there, they're getting away!

I'm not sure I'm becoming a proper fisherman. I've got a feeling I'm becoming someone who really likes to go out into the British countryside and escape. That's now obvious, but somehow in the previous 30 years I hardly ever did it at all.

I've started going out to the countryside more with the wife as well, because it's very worthwhile, although I can't pin it down. Being away from home is nice. And I also enjoy coming back home. You know, a little pensioner adventure. A lot of pensioners meander around garden centres now, don't they? And coffee shops at castles. I hope I'll be a pensioner who meanders around riverbanks. It's a grand day out. I'm becoming a true devotee of nice pensioner days out.

So fishing's not just about catching a fish. It's about a lot of things, and when they all come together, it's unbeatable.

I have an overpowering feeling when I've had a day fishing that I've used the day well. And you can get a bit obsessive about it. Maybe now I'm down to – I don't know what it is – 3,000 days left on this earth? Of the last 20, I used that one day really well. That's what I feel.

Bob: There is a little gift element to a serious ailment, isn't there? A side benefit. I used to look at my mother-in-law when she came round and she'd stare at birds. Wouldn't say owt but she'd stare at the birds. It was boring for us – we'd be trying to watch the football and she'd say, 'Oh, look at the birds on that tree.' And thinking you might die, you get a little window into that mindset. Being able to be alive and watch a magpie *is* incredible, and you ignore it. So I got a little insight – I understood exactly where she was at that point in her life. She didn't have long left at that time, and when something like this happens, you realise you have to drink from it while you can.

Paul: It's like the opening line of TS Eliot's *The Waste Land*. 'April is the cruellest month.' Because up come the daffodils and all the flowers again, and you'd better look at that beauty, because you ain't getting to see it again.

B: Look at this beauty that you have completely and totally ignored for 30 years. You've more or less ignored it. If you had taken me fishing 20 years ago, it wouldn't be half the experience it is now.

P: And that's exactly what it is – an experience. My dad used to take his mate's son fishing and I'd go with them sometimes. My dad's friend had a serious illness that left him confined to a wheelchair for much of the time. He was certainly unable to walk up a wild riverbank. So one beautiful spring morning on the River Usk, my dad and his mate's son left him on a bench by the river and went off fishing for a couple of hours. When they returned and asked him how he'd been, he said it was the nearest thing to paradise he'd ever experienced in his life. His 'confinement' (for want of a better term) had forced him to appreciate the colossal beauty in front of him that most of the time we don't even acknowledge, let alone appreciate.

And it's true: once you're out there, you do start to praise even silly little things. We've had a rainbow come out a couple of times when we've been fishing and you do go, 'It's a sign!' And it is! It *is* a sign! It's saying, 'Look how *beautiful* life is!'

CHAPTER 5
A BRIEF HISTORY OF FISHING

Bob: Our friend Izaak Walton checks the football scores on the banks of the River Lea.

'If ever you have an evening to spare, up the river, I should advise you to drop into one of the little village inns, and take a seat in the tap-room. You will be nearly sure to meet one or two old rod-men, sipping their toddy there, and they will tell you enough fishy stories, in half an hour, to give you indigestion for a month.'

Jerome K Jerome, Three Men In A Boat *(1889)*

Paul: Much like Izaak Walton's *The Compleat Angler*, I thought we could deliver this chapter in the style of a learned fisherman talking to a friend who is ignorant but keen to learn about the history and lore of fishing.

Bob: Which role do you wish me to take, Paul?

P: Ideally the fisherman, because it would mean a lot less work for me.

B: No, but you're good at it, Paul. You kick off, I'll interject.

P: Well, let us begin at the dawning of time.

B: This chapter sounds like it's going to be quite long.

P: It is. History has not recorded the name of the world's first fisherman. But one fact about him is guaranteed. Five minutes after this unknown genius popped a hook on a stick and slung it in the water, I have no doubt, Bob, that a random bloke undoubtedly wandered over, stood a few metres behind him and told him he'd never catch a fish in this bit of the river using that tackle.

Then would have followed the first riverbank conversation, one which would be repeated almost identically over the coming centuries and millennia. These two men would have quickly agreed that the conditions were hopeless: either too hot, too cold, too windy, no breeze, wrong wind direction, no cloud cover, river too low, river too high, wrong time of year, wrong barometric pressure, wrong bait … you name it. And they would have left each other, little aware of their place in history, but both thinking the other one was full of it.

Man has been fishing for as long as he's been doing anything of note. Archaeological digs have established that the first fishermen were *Homo habilis* 1 million years ago, although they were largely monkey-men and probably just using their big, hairy monkey hands to grab fish.

B: Blimey, Paul, considering it can be bad enough when you've got all the kit, what a terrible day's fishing those lads must have had! I'd have not bothered, just stuck to the berries and grubs. Maybe that's why they didn't survive to the modern day, because they were stuck on the riverbank all day, catching nowt.

P: Well, *Homo habilis* were followed as fisher-monkey-men by *Homo erectus* and then *Homo sapiens*, who may even have used some makeshift fishing kit, although none of it has survived. With your simple blue rod, Bob, you're a throwback to the days of the monkey-fisher.

The first identified *Homo sapiens* fisherman was Tianyuan Man, whose bone fragments – a bit of lower jaw, a leg bone – were discovered in 2007 in a cave near Beijing. Radiocarbon dating showed his bones to be around 40,000 years old, while isotope analysis of his bones showed a substantial part of his diet came from eating freshwater fish. Mind you, archaeological digs can be misleading: for example, I put my dad's ashes into the River Usk in Wales, the Test in Hampshire and the Dee in Scotland. As he had the ashes of two of his dogs mixed in with his, any archaeological bone/gene study would conclude there was once a species of dog-men that lived in various settlements around England, Wales and Scotland.

B: That's very true, Paul. Did further evidence of Tianyuan Man being a fisherman emerge the following year, when the bones of his wife were discovered, showing that she'd been left alone all Saturday, even though he'd promised he'd be back at six because he knew they were going out and it had been on the calendar for weeks?

P: An interesting theory, Dr Mortimer. Anyway, around 3500 BC, at the time the Ancient Britons were discovering agriculture, the spear net, line and rod appeared almost simultaneously in Egypt. Five hundred years later, the Egyptians were using baited hooks made of ivory, bone and shell dangled off the side of boats and dragged down with clay weights.

In short, no other pastime has been as well documented throughout ancient history as fishing. In many ways, the history of fishing is the history of the tackle that we used – after all, the fish haven't changed, the goal is still exactly the same (to catch a fish successfully) and nor has the mindset of the fisher. Whether they're Stone Age men or a modern Hoxton idiot with a £200 neon haircut, despite being separated by millennia, they're doing exactly the same thing: sitting out on the riverbank, staring at the end of the rod and waiting for it to show the slightest quiver from the blubbery lips of a fish. It's the greatest puzzle that nature ever set.

B: Fishing is also the only one of today's popular leisure activities to make it into the Bible – if you have a read of the Good Book, Paul, you'll see none of the disciples are into jogging, the Israelites don't go into the desert to play bridge, and Christ performs no miracles with a paintball gun. It's fishing all the way.

P: And the Biblical fish references are endless. The disciples are fishermen; 5,000 people are fed with two fish; Jesus is called a fisher of men; and one of the Three Wise Men turns up with a massive bowl of smoked trout pâté (this may or may not be true).

Anyway, the earliest mention of fishing as a sport – rather than just a way to stave off starvation for a short while – comes in 1496 and surprisingly, the book it's in is as 'woke' as fishing gets. *A Treatyse of Fysshynge wyth an Angle* was the first English book to give details about making and using a fishing rod – and was surprisingly written by a woman, Dame Juliana Berners.

Not much is known about Dame Juliana; in fact, there's so little information about her, some people don't believe she actually existed, let alone authored a key book on fishing. But *A Treatyse of Fysshynge wyth an Angle* (at the time, 'angle' meant a hook – this is why we're called anglers instead of hookers) has survived for 500 years and while every fisherman has heard of the book, very few of them have actually read it. Most of the reprints tend to be facsimiles of the earliest editions and if your Middle English is rusty, Bob, it can a bit of a slog. Here's one of the first sentences:

> The beste to my symple dyscrecōn whyche is fysshynge: callyd Anglynge wyth a rodde: and a lyne and an hoke.

Sure, Dame Juliana might spell fishing 'fysshynge' (which if nothing else would make her a tremendous guest on *Countdown*'s Dictionary Corner one week), but what's remarkable is how her 15th-century attitude to angling is virtually the same as the modern fisher's. For example, nothing is more timeless than her list of the 12 impediments that stop you catching a fish.

1. the first is if your tackle is not adequate nor suitably made;
2. the second is if your baits are not good or fine;
3. the third is if you do not angle in biting time;
4. the fourth is if the fish are frightened by the sight of a man;
5. the fifth, if the water is very thick, white or red from any recent flood;
6. the sixth, if the fish cannot stir because of the cold;
7. the seventh, if the weather is hot;
8. the eighth, if it rains;
9. the ninth, if it hails or snow falls;
10. the tenth is if there is a tempest;
11. the cleventh is if there is a great wind;
12. the twelfth if the wind is in the east, and that is worst, for commonly, both winter and summer, the fish will

not bite then. The west and north winds are good, but the south is best.

13. the thirteenth is if Robert Mortimer is with you.

B: I love this, Paul. That list is the foundation of fishing excuses as they're still practised today. You still hear those exact excuses trotted out every time you meet another fisherman on the bank. It's not a bad bit of observational comedy considering the author was a medieval nun who might not have actually existed.

P: Well, much of the early modern history of fishing is known only from books, for the simple reason that the physical tackle hasn't survived – it was made of wood and hair and used until it was broken or simply rotted. Which brings us to the most celebrated early book about fishing: *The Compleat Angler* by Izaak Walton.

B: Ah yes, the world's fourth most reprinted book (it says here), beaten only by the Bible, the works of Shakespeare and the Book of Common Prayer.

P: No surprise there at all. It's an absolute gem of a book and every fisherman's friend. Here's why: born in Staffordshire in 1593, Walton worked as a linen merchant in London, but returned frequently to his home county to visit friends, who more often than not would take him fishing. Following the Civil War, Walton retired from London and moved back to

Staffordshire, where he bought a little timbered cottage farm in the tiny hamlet of Shallowford. It backed onto a winding brook called the Meece (a tributary of the Sow) and Walton spent the rest of his life wandering off to meet friends who enjoyed fishing, and slowly began compiling what he learned from them into a book.

Walton's cottage still stands – it's no longer adjacent to the little brook as the Grand Junction Railway was built over his garden in 1837 – and it's now home to a museum dedicated to Walton's varied career, along with a room dedicated to the history and evolution of angling.

B: I've had a look on TripAdvisor, Paul, and there's not many reviews of the museum. The most recent calls Izaak Walton's Cottage 'a great place to take children to learn about this great man'. It's also the *only* place where you can take children to learn about this great man, but let's not split hairs. 'Not a full day out,' says another person, but you could just do it slowly and it would be.

Another reviewer left two separate reviews banging on about how good the ice cream was – she said it was 'lovely' in October 2016 and then seven months later, she returned and found it 'adorable'. So if you do go, make sure you get an ice cream; they must be knockout.

P: Or go and see his tomb in Winchester Cathedral. Plenty of ice cream round there. Anyway, back to the story. In 1653, at the age of 60, Walton published his book about fishing. The book begins with three strangers – Piscator (Fisherman), Venator (Hunter) and Auceps (Falconer) – meeting as they travel north of London from Tottenham (COYS!), through the valley of the River Lea towards Ware in Hertfordshire. Auceps soon goes his own way, but as the other two continue, the older Piscator teaches the younger Venator how to become an angler; just as we two are doing in the present day, albeit with more swearing.

The book was an immediate success, and Walton spent the rest of his life – and he lived to 90, a really uncommon age in the 17th century – revising it and issuing new editions. In the years after its publication, he spent a lot of time fishing on the Dove with his friend Charles Cotton, who contributed a section on fly fishing for the fifth edition (and the last to be published during Walton's lifetime).

Mirroring the characters in the book, Cotton had been a keen hunter, until Walton 'taught me as good, a more quiet, innocent, and less dangerous diversion'. Cotton eventually frittered away his family fortune on wine, women and trout, but made good in the end by shacking up with a wealthy widow, and was able to fish until the end of his days. Cotton's axiom, 'Fish fine, and far away' remains a fundamental rule of angling.

It's worth saying that, while it's a brilliant read and still thoroughly enjoyable from start to finish, there's quite a lot in *The Compleat Angler* that seems absolutely mental today. For a start, Walton doesn't write very much about *actual fishing*. In the opening address to his readers, he says there's no point in him explaining how to fish, because it won't be of any help to anyone. So instead of being a guide to fishing, the book is largely 200 pages of Walton continually getting distracted by the wonders of the British countryside. So there's a whole chapter about the joy of killing otters ('Kill her merrily!'), as well as tales of the randy Sargus fish (which he believed 'goes courting she-goats on the grassy shore'); there's also Walton's belief that eels grew from weeds, and about five pages of songs about milkmaids.

B: Didn't Walton also explain how to fish for pike by tying bait to a live duck's leg and chucking it in the water or something? The idea is the pike tries to bite the duck's legs off, the two get tangled together and then you can drag the whole screaming lot in. It doesn't sound like the most relaxing fishing you'll ever do. Or the most politically correct.

P: No, but much of what Walton writes about still rings true today, and the delights of fishing and the British countryside have never been better conveyed. Walton created the perfect vision of what fishing should be, and his vanished, serene Arcadia is a folk memory that every British fisherman has tangled up in their DNA:

> Under that broad beech tree I sat down. There I sat viewing the silver streams glide silently towards their centre,

the tempestuous sea; yet sometimes opposed by rugged roots and pebble-stones, which broke their waves and turned them into foam. And sometimes I beguiled time by viewing the harmless lambs; some leaping securely in the cool shade, whilst others sported themselves in the cheerful sun; and as I thus sat, these and other sights had so fully possessed my soul with content, that I thought, as the poet hath happily expressed it,

I was for that time lifted above earth,
And possess'd joys not promised in my birth.

B: Lovely.

Silence.

P: So, fishing (and the rural way of life in general) changed little from Walton's time until the Victorian era. There were some minor developments – the wooden rod with a single horsehair with a bit of bait tied on the end were upgraded with little metal loops, allowing the use of a running line (an advance which ultimately led to the invention of the fishing reel).

The first mention of a reel comes in 1651's *The Art of Angling*, a similar time to when metal fish hooks started to be developed. In the middle of the 17th century, Charles Kirby invented a hook with an offset point, which he began manufacturing in Redditch. The Kirby Hook is still in use across the world today, and Redditch remains the centre of the hook-making industry in England.

It was in the Victorian era that fishing changed dramatically. Angling morphed from being something working men pottered off to do during a spare hour, in the hope of hooking something for the dinner table, to becoming the celebrated and rarefied sporting preserve of the elite.

In the mid-19th century, the wealthy started escaping from the soot, grime, infection and overcrowding of the newly industrialised cities and began buying vast tracts of unspoiled land in the peaceful, wholesome countryside, well away from the unwashed masses and the horrible factories that made them their fortunes. These huge estates didn't just comprise mansions, gardens, woodland and meadows, but frequently came with the ownership of the rivers that ran through them. Over a matter of decades, countless rivers fell into private hands: at one point, the entire 107-mile length of the Spey was owned by just six holders of vast estates. It meant that for the first time in history, rivers became inaccessible to the majority of the ordinary people who lived close to their banks.

Seduced by the suddenly fashionable countryside, the Victorian aristocracy quickly became fanatical about fishing and the cachet of summer fishing retreats was firmly cemented in 1848 when Queen Victoria leased the Balmoral estate, buying it outright in 1852. In the days before international travel was practical, this led to summer hunting and fishing trips to Scotland became a key part of the aristocratic social calendar.

With the upper part of the Dee Valley running through the estate, fishing was granted the Royal seal of approval.

In September 1850, Victoria drew a sketch of the estate workers fishing for salmon with poles and spears on the Dee, and she is popularly believed to have enjoyed a close friendship with the late Prince Albert's personal gillie, John Brown. In a personal diary entry, she said of Brown: 'He is so devoted to me – so simple, so intelligent, so unlike an ordinary servant, and so cheerful and attentive' and remarked on his 'strong, powerful arm'. That might not sound explosive, but this is the emotionally repressed Victorian equivalent of sexting.

One of Victoria's daughters, the rebellious Princess Louise, Duchess of Argyll, was also renowned as a decent salmon fisher in her day. Princess Louise toured Canada in 1879 – it was billed as a state visit but actually it was arranged because she and her husband, John Campbell, 9th Duke of Argyll wanted to go fishing in the Cascapedia River. While there, she caught a 28lb Atlantic salmon, causing her husband to remark that fishing in Canada took no skill. Princess Louise's beloved eight-foot lightweight greenheart and hickory salmon rod, made by John Forrest of Kelso, came up for auction in London in 2001.

B: Yeah, and hopefully, that very rod was taken by Princess Louise and shoved up her husband's arse the minute after he made that weak joke about women not being able to fish, the rude, misogynistic sod!

P: That goes without saying and is very 'right on' for you. But as we know, Bob, the Royal Family still own Balmoral and they still fish the Dee. The *Shooting Times* reported in 1953 that the Queen had taken up angling seriously after her marriage, and 'since then, she has developed a real love for it'. Philip's enthusiasm for fishing has also grown, especially since he had to give up shooting after having a stent fitted in 2011 (not that stents stopped me for one second. Just saying!). The *Sun* reported in 2018 that the 97-year-old still 'sometimes stands for hours in the River Dee', only coming out of the water 'to join the Queen and guests for lunch or tea'; no doubt desperate to insult someone after all that time alone and probably fishless. They don't catch many salmon up at Balmoral these days.

B: But what about the Queen Mother, Paul? Wasn't she the Royal who loved fishing the most, Gawd bless 'er? And didn't she almost choke to death on a fish bone at a Balmoral banquet in 1993? I imagine the Queen Mum is now up in heaven, fishing merrily in a lake made out of gin and tonic.

P: Well, when it comes to eating fish, the Queen famously avoids shellfish when she's out and about (not to be overly delicate about it, but Her Majesty doesn't want to get the quicklies if it's gone over the top), but her personal cook revealed she often eats grilled Dover sole for lunch and then salmon from Balmoral in the evening.

Also, did you know, Bob, that in one of the strange quirks of English law, whales, porpoises, dolphins and sturgeons (whose eggs are famously extracted as caviar) are designated as 'Royal fish', and when caught in Britain, they immediately become the personal property of the monarch? This is why when that whale died in the Thames a few years ago, the Queen could be seen trying to drag it into the boot of her car and telling passers-by to 'jog on and get your own dead whale, you mugs'.

B: But this whole Royal fish nonsense doesn't apply today, does it?

P: Yes, Bob, yes, it does. You have Edward II to blame for that. Well, a chap named Robert Davies does, at the very least: in 2004, the Welsh fisherman caught a 264lb sturgeon in Swansea Bay and sold it in an auction for £650. He was then arrested and the fish impounded. Knowing it's not illegal to catch or sell a sturgeon so long as you offer it to the Queen first, Davies said he'd faxed Buckingham Palace and they'd replied, telling him to keep it.

B: Incredible – a fax machine still being used in 2004; who'd have thought?

P: The police then informed Davies that they wouldn't have got involved if he'd kept it or given it away free – instead, they threatened him with six months in prison. *This country!* (Incidentally, I had a communist colleague in Hackney Council who reckoned fishing was counter-revolutionary because all classes did it! He was joking but there was a definite Stalinist threat behind it.)

Anyway, back to the Victorians. With the rich embracing the sport, fishing swiftly became a huge and lucrative business, with manufacturing companies springing up across Britain: reels began to be mass-produced, lighter, more flexible rods were made with wood and bamboo imported from the empire, and horsehair gave way to stronger silk lines, which were lighter and enabled them to be cast further.

In 1874, two Northumberland brothers, William and John James, set up their company, Hardy Brothers, to manufacture fishing rods, gaining a Royal Warrant in 1891 and opening a famous shop at 61 Pall Mall, London, which lasted until the early 2000s. Of course, history remembers it primarily as the place where the comedian Paul Whitehouse helped the comedian Bob Mortimer pick out a rod in the years running up to their BAFTA-winning (pending at time of writing) TV show *Mortimer & Whitehouse: Gone Fishing* (not that Paul needs any more BAFTAs – he already has five (yes, Bob, *FIVE*).

B: Ah yes, that was a lovely day, Paul.

P: It was indeed, Robert. A shame the shop isn't still there now …

Like a lot of Victorian crazes, the fishing mania eventually calmed and settled, although it left Scotland with a legacy of tremendous fishing lodges and fisheries that are still there and in use today.

But the development of fishing tackle continued apace. In 1905, the textile magnate Alfred Holden Illingworth filed the first patent on the fixed spool or spinning reel. Daydreaming about fishing in his factory one day, he noticed the shuttles going backwards and forwards on one of his vast weaving machines and realised something similar could be used to stop his silk line getting tangled when he cast it. His invention meant casting distances increased dramatically and revolutionised freshwater fishing. Every new type reel made since – and there's been *millions* of them, let me tell you – simply builds and improves on Illingworth's original barnstormer of an idea.

By the 1920s, as tackle was refined and perfected, some of the best writing on fishing came from the pen of Arthur Ransome. Best known for the children's book *Swallows and Amazons*, Ransome worked as a Russian correspondent during the First World War and was on nodding terms with Lenin, Trotsky and MI5 (for whom he worked as a spy).

Ransome had been an avid fisherman since the age of ten, when his postman took him for a day out at the Wey outside Godalming and he landed two roach. During the 1920s, Ransome wrote a regular fishing column for the *Manchester Guardian*, and pretty much everything in those is brilliantly quotable.

Fishing's next great boom period came at the end of the Second World War. The development of travel – the railway network, coaches, cars and motorbikes – meant city dwellers could head out to the country more easily than ever before. In 1953, *Angling Times* launched and quickly became Britain's largest angling newspaper. Perhaps its biggest legacy is popularising the classic 'fisherman-kneeling-and-holding-a-big-fish-while-smiling' photograph, which you can still see on every one of its covers today. In fishing circles, getting a photo in *Angling Times* of you holding a good-sized tench is the equivalent of being a supermodel who gets their first *Vogue* cover.

B: Yep, and in 2018, *Angling Times* named *Mortimer & Whitehouse: Gone Fishing* their TV show of the year, and honestly, that accolade is the highest honour we could ever have wished for.

P: Absolutely, Bob. But obviously, we're not the first fishing show on mainstream TV. While fishing's not been particularly well served by television over the years, there have been some absolutely belting series over the last decades. In the 1980s, Anglia TV's *Go Fishing* saw John Wilson, the owner of a tackle shop in Norfolk, head out to catch some fish. In 2004, the readers of *Angling Times* voted John the greatest fisherman of all time. Sadly, he died in November 2018 – we heard while filming *Gone Fishing*, fruitlessly attempting to catch grayling in the Ure, and it came as an especially bitter blow for our fishing consultant, John Bailey, who was long-standing friends with John. There are a lot of his shows on YouTube, and all of them are a lovely tonic.

In 1993, the fantastic *A Passion For Angling* was shown on a primetime Sunday evening slot on the BBC. Beautifully filmed over four and a half years, the six episodes saw Chris Yates and Bob James go fishing across Britain. They believed that 'the secret of success is not how to catch, but how to enjoy'.

B: Sounds familiar … I'd say my fondest memory of that show is when a barn owl nearly landed on Chris's hat. I would *love* that to happen in our show.

P: Well, we nearly had that swan land on you, and you had some interest from those cows at the Wye, didn't you, Bob? The memories …

Anyway, that brings us to today. In 2018, recreational fishing looks to be in cautious good health. Listen carefully, Bob, because I'm going to reel off a few stats:

- It's estimated 2.2 million people still go coarse fishing, with another 800,000 game anglers, and between them they contribute an estimated £1.4billion into the English economy through their spending on the sport.
- There are hundreds of thriving angling clubs and associations.

- The 42,123km of rivers in the UK have never been looked after by more people with an eye to preserving and protecting them.

B: That's great to hear. And the readers can find out about our favourite locations a little further into the book, can't they, Paul?

P: That they can do. But it should be said that as the landed gentry have found the upkeep of their vast estates and hire of manservants financially draining, most of the country's riverbanks have been opened up for day fishing, meaning anyone can fish on the country's iconic rivers for a modest outlay of cash – most of which goes straight back into the management of the river and the immediate environment.

But the most heart-warming sign about the future of fishing comes from a huge 2018 survey, where the Environment Agency found that 'a clean and attractive environment with minimal disturbance was more highly valued by anglers than the size and abundance of fish'.

It just goes to show that even though fishing has evolved greatly due to technological advances over the centuries, very little seems to have changed in the mindset of the angler since Izaak Walton was sitting on the banks of the Lea, 350 years ago:

> Let me tell you, there be many that have forty times our estates, that would give the greatest part of it to be healthful and cheerful like us; who, with the expense of a

little money, have eat and drank, and laught, and angled, and sung, and slept securely; and rose next day, and cast away care, and sung and laught, and angled again; which are blessings rich men cannot purchase with all their money.

CHAPTER 6

A HEALTHY FISHERMAN'S RECIPE COLLECTION

BY ROBERT MORTIMER

Bob: Did someone say 'pie'?

Bob: Do you know what? All that talk of fishing has sparked a Pavlovian response in me. I'm ready to have some nice food. I've had a proper run of being heart-healthy recently. Genuinely, Paul – I've done nuts, crackers, yoghurt, bananas.

Paul: Well done!

B: Although I did have a pie, Paul. One pie. I justified it because it was in the fridge and it was getting close to where it needed to be thrown out – so I had it.

P: Look, we're all a bit bad sometimes. Everything in moderation, including moderation. Even the surgeon echoed that.

B: You do sort of have that mind game where you think to yourself, 'Look, I've been out on the river all day, so I deserve to have a pie.'

P: And that's fair enough. Did you enjoy it?

B: No. It was a sad pie, just a sad little pie.

P: Why didn't you make yourself one of your heart-healthy recipes? They're really good. Much better than anyone expected.

B: I should have, Paul. I should have made myself the world-famous Bubbles McParty's sardine and feta cheese wrap. That's the best one.

P: Is it better than a sad little pie?

B: Much better than a sad little pie.

BOB

Once you have been diagnosed with heart disease, food takes on far greater importance in your life. Eat the wrong foods and your arteries will start to clog up again. Food is no longer simply for sustenance and pleasure, it becomes a matter of life and death.

You can, of course, choose to eat anything you want, but further down the line it will destroy you. I imagine some people do the calculations and decide that they would rather have ten years of indulgence than 20 years living off seeds and fruits. For my part I try to strike a balance between pleasure and pain. It may take a few years off my life but is a life without cheese and pie really worth living?

The big bad wolf is saturated fat, and basically, any variety of animal fat. So no butter, no cheese, no meat, no ice cream, no fried food, no bacon, no ham, no buttery cakes or biscuits. NO PIES! Basically, none of the things that I, and before me my mum, had shovelled down my neck for the previous 50 years. I was addicted to this stuff and my guess was that it would probably take another 50 years to wean myself off it. Cheese, I knew, would be a big problem.

The last medical professional I saw before leaving hospital following my surgery was a heart dietician. She waxed lyrical about the joys of oatcakes and vegetable soups; the intense gratifying pleasure to be found inside a tofu lump and the finger-licking goodness of a cauliflower fritter.

I asked her about cheese and meat. 'No cheese, no meat,' she replied. I tried a different tack: 'What if I only eat seeds, nuts and berries for the whole week? Could I have a nibble on some cheese on the Sunday evening?' 'No cheese, no meat.'

I told her that I didn't think 'no cheese' was going to work for me: I'm a very cheesy bloke. She realised I was 'cheese weak' and softened her stance slightly: 'You can have one matchbox-sized portion per month.'

I, of course, took this as a household-size box of matches and felt I'd won a small victory. I didn't mention my interpretation to the nurse.

For the first six months after the operation I was a very good boy. Skimmed milk in my tea. Olive oil spread on my toast.*

* Paul: Some say butter is good for you. Basically, somebody is out there contradicting everything.

Porridge for breakfast. Pears and plums as a snack. Eggs or sardines for lunch and vegetables for tea. I found a brand of potato crisps cooked in olive oil and that became my nightly treat. At the weekend I'd eat a lump of cheese shaved thinly with a potato peeler so that I could have it on crackers and make it last until Sunday night.

Since then I have slowly reintroduced more and more treats into my diet, but that's what they are … treats. I try hard to avoid saturated fats and whenever I am tempted, I think of those huge fat balls that block our sewer pipes … it's usually enough to persuade me to behave.

Eating fish is one of the great culinary pleasures I can still enjoy (but not shellfish, which is very high in cholesterol). The first fish I ever caught was a small brown trout on the River Esk at Glaisdale in Yorkshire. I cooked it on the riverbank using a small fire made from lolly sticks and twigs. I can still taste it now. Wonderful.

It is one of the great joys in life to cook by an isolated river or lake but it is something that few of us ever do. When Paul and I were planning the filming of *Mortimer & Whitehouse: Gone Fishing* I was desperate to be the chef and for us to film the cooking and the eating. 'Nobody wants to see a clown like you cooking,' they said, but I always thought it would resonate with the thousands of amateur cooks who have struggled to knock up something edible on a camping stove with ice-cold fingers and only a plastic spork for cutlery.

When two people on a heart-healthy diet get together, conversation can very quickly turn to food and a process of euphoric recollection commences. Sometimes as Paul and I sat watching our floats, a conversation would emerge out of the silence:

P: Pie?

B: Which pie?

Silence.

P: Steak.

B: With kidney?

P: No.

Silence.

B: Shortcrust or puff?

P: Shortcrust, you puff.

Silence.

B: Gravy?

P: Yes.

B: On pie or near pie?

P: In jug.

B: Then poured on pie or near pie?

P: Tricky.

Long silence.

P: On pie.

B: Nice.

During the filming I tried to reflect my 'fairly heart-healthy' diet. Always a heart-healthy breakfast, a 95% healthy bankside snack and a 90% healthy supper. Then a real couple of treats per series.

When it comes to those treats, or foods that I miss terribly, I would list them in the following order of sadness for their loss:

- Cheese – I now have a tiny portion once a week
- Meat pies – maybe three times a year
- Cream cakes – once a year
- Bacon – once a month
- Sirloin steak – three times a year
- Sausages – two sausages every fortnight (it's the law in my house)
- Roast dinner – six a year

- Syrup sponge pudding and custard – four a year
- Doner kebab – I dream about them every Friday night
- Chocolate – unwilling to disclose average consumption.

So tot it all up and you will see that there is always a treat on my event horizon and of course absence makes the heart grow fonder.

The rest of my diet is the usual low-fat staples – oats, vegetables, nuts, pulses, yoghurt, bread, eggs, etc., etc. I must give a special mention to a little group of foods that I try to eat some of every single day – chestnuts, pears, plums, prunes and dates. My understanding is that they contain a particular type of fibre that the body needs to use bad cholesterol to digest. I tried to incorporate them in lots of the dishes I prepared for Paul during *Mortimer & Whitehouse: Gone Fishing*. I think he's now a believer, but having said that, he will say anything to shut me up.

Here are some of the recipes I have cooked during the series. The sardine wrap is my particular favourite, though if you were to ask me which meal I enjoyed most during the show it would be the steak pie we ate in the Norfolk pub after I beat Paul in the roach competition – another meal I will never forget.

THE RECIPES

The key to bankside cooking is undoubtedly minimum ingredients and preparation time. All the ingredients and utensils for the meals I cooked in the series were easily transported in one small cool box.

TUNA MELANIE WITH TRAPPED POTATOES

I cooked this dish for Paul on the banks of the majestic River Wye at Monsal Dale in Derbyshire. It is undoubtedly one of the most beautiful places on this earth – hobbit country. The recipe only requires one frying pan, two plates and a spoon and fork, or a spork!

Tuna is very low in saturated fat and high in omega-3 fatty acids, which help increase levels of HDL (good) cholesterol. It is also amongst the tastiest of the ocean meats.

Tinned, or trapped, potatoes are one of my favourite foods. The surface of the spuds is firm but absorbent, allowing it to soak up flavour without hesitation. I always fry them in olive oil and add a pertinent herb. In the case of tuna, thyme is a great enhancer. Don't be afraid to let the potatoes crisp up a bit on the outside – it adds a nice bit of texture. The tuna and the spuds are mysteriously blended into perfect harmony by the addition of marmalade. You won't believe me until you give it a try – and I hope that you do.

Ingredients (serves two old farts):

3 tsp olive oil

A pinch of pepper

Dried thyme

1 tin of potatoes

2 x 125g/4oz tuna steaks, drained

2 tsp marmalade

Heat the oil, pepper and thyme over a medium heat. Add the potatoes and fry until they just begin to brown. Add the tuna steaks and put a good-sized teaspoonful of marmalade on top of each one. Turn after two minutes and again, place a dollop of marmalade on top of each steak. Cook to your liking and serve.

PARK AND RIDE ONE-PAN CHICKEN PILAF

Paul and I ate this on the banks of the beautiful River Wensum in Norfolk. It was a sunny, late autumn day with a real chill in the wind. It was a pleasure to warm my hands by the stove and watch on as Paul failed and failed again to land himself that elusive trophy roach.

Lean chicken breast is very low in cholesterol and the inclusion of dates provides some cholesterol-fighting fibre. The dish only takes 20 minutes to prepare and cook. I don't know how long it takes to clean up afterwards because that was Paul's job and I never saw him do it.

Ingredients (serves two buffoons):

1 tsp olive oil

1 small onion peeled and chopped

1 diced skinless chicken breast

1 tsp curry powder or paste

A third of a mug of basmati rice

Two-thirds of a mug of chicken stock

1 mug of frozen mixed vegetables

A handful of dates

Heat the oil in a non-stick pan, then fry the onion until soft. Add the chicken pieces and fry for a further couple of minutes just to colour the outside, then stir in the curry powder or paste and rice. Cook for another minute.

Pour in the chicken stock and stir in any larger bits of frozen vegetables. Bring to the boil, lower the heat, then cover the pan with a lid. Cook for ten minutes, then stir in the remaining vegetables and the dates and cook until all the stock is absorbed.

Spork it down your neck.

BUBBLES MCPARTY'S SARDINE AND FETA CHEESE WRAP

These little pleasure packets are best pre-prepared and wrapped in foil. Carry them with you as a pocket snack for whenever the need arises. We had them in Norfolk by a lake, float fishing for tench and then again, at Paul's request, at Throop Mill pond when ledger fishing for pike.

Feta cheese is on the lower end of the cheese scale for saturated fat but is still, strictly speaking, illegal for a heart-healthy diet. If this is unacceptable then substitute the feta for a low-fat cottage cheese. Sardines are high in omega-3 oils to boost your good cholesterol. It is the addition of fresh mint and dates that elevates this simple wrap to the Premier League of pocket meat treats.

Ingredients: (serves two Herberts)
1 pitta bread
1 tin of sardines in lemon oil
50g/2oz feta cheese
Tomato chopped
Lettuce chopped
Dates pitted
Fresh mint leaves
Coriander

Open up the pitta and thrust the remaining ingredients inside with partial fury but a hint of love and care. Use your historic opinions of the ingredients to judge your ratios. This dish crept up on Paul and gave him quite a startle. It's a bigger hitter than you may think.

SLIPPERY WILF'S CAULIFLOWER BASE PIZZA

An alternative to a bread-based pizza if you are trying to cut down on the carbs. Put any topping on apart from pineapple which, in a savoury context, should only ever be heated when it is on top of a thick slice of gammon. Paul and I ate this in our fishing lodge next to the Herefordshire Wye. We had just caught two magnificent barbel in the stretch adjacent to the lodge and could not have been happier. After eating the pizza I was moved to strip off and jump in the hot tub … maybe it will have a similar effect on you.

Ingredients (serves two laughingstocksmen):
500g/1lb cauliflower rice
1 egg
Half a cup of finely grated Parmesan cheese
Dried oregano
Dried basil
Salt and pepper
Tomato purée
Topping of choice

Place the cauliflower rice in a bowl and microwave on full power for three minutes. Pour the rice onto a tea towel and fold the towel around the rice. Then twist the towel with all your might to force all the moisture out of the rice. You will look magnificent during this process.

Put the post-squeezed rice back into a bowl and add the egg, Parmesan and seasonings. Using your beautiful hands, spread the mixture into a pleasing, potentially circular, shape on a sheet of greaseproof paper. Place in a hot (200°C/Gas 6) oven for five to six minutes until it just begins to brown. Remove from the oven and add the topping of your choice.

NOT NOW MADAM VENISON, CABBAGE AND CHESTNUT

This was actually the first meal I cooked for Paul when we filmed *Mortimer & Whitehouse: Gone Fishing.* We were staying in a converted water tank on top of a hill in the Derbyshire Dales. At this stage he was curious about my cooking skills but not full of expectation.

I burnt my hand quite badly when cooking the cabbage for this meal. Rather than help me, Paul chose to save the cabbage. What a hero.

Venison is a great meat for heart-healthers – very low in saturated fats and cholesterol. Of course you would be better off eating a root or a block of ice but if you've got the hots for a piece of meat then venison is the best choice. The addition of chestnuts gives a bit of cholesterol-busting power to the dish and a nuttiness that goes great with the venison and the cabbage.

Ingredients (serves two simpletons):

1 Savoy cabbage

Nutmeg

1 tbsp olive oil
salt and pepper
2 x 125g/4oz lean venison steaks
125g/4oz vacuum-packed chestnuts, roughly diced
One slug of red wine

Remove the outer leaves and core of the cabbage and slice thinly. Place in a saucepan with 1cm of water and a good pinch of grated nutmeg. Boil/steam for four to five minutes.

Heat the olive oil in frying pan. Season the steaks and fry with the chestnuts for two to three minutes each side (or according to preference of doneness).

Drain the cabbage and add to the frying pan with a slug of red wine. Cook off the wine and serve.

BOOTS MCFOOLISH BREAKFAST PORRIDGE

I cooked this for Paul's breakfast most mornings. It's a powerful start to the day and gives you that glowing 'I just done right by my heart pipes' feeling. Oats, prunes, pears and plums in a big cholesterol-baiting bowl.

I top my porridge with brown sugar. Paul tops his with chia seeds, because he's from London.

Ingredients (serves two halfwits):
300ml/½ pint semi-skimmed milk
300ml/½ pint water

175g/6oz rolled oats
Large pinch of salt
A few tinned prunes
A handful tinned pears
A handful tinned plums
Brown sugar or chia seeds

Put the milk, water, oats and salt into a saucepan, stir and bring to the boil. Reduce the heat and simmer slowly for seven to eight minutes. If you can be bothered, the more you stir the porridge as it simmers, the creamier it will be.

When cooked, remove from the heat and fold in the fruit according to your ratio preferences. Serve in a bowl and sprinkle brown sugar on top, unless you are from London: in which case feel free to use chia seeds or mouse whiskers.

B: The sardine sandwich is the best thing I cook, Paul. It's delicious.

P: It is, but this isn't just a book about going to the supermarket and tipping some fish into a pan, Bob – we're supposed to be catching them.

B: I think out of all the fish I've caught, the ones I like most are the tench or the barbel. It took a long time to catch the barbel, but the tench … It was a beautiful place, really quiet, and everyone had gone and left us and BANG! Here's your tench, sir.

P: It was a lovely fish and a lovely bit of fishing. I've also got a lot of affection for the wise old tench.

B: I'd accept the tench as the wisest of the fish. But on that afternoon, it was no match for you, was it, Paul?

CHAPTER 7

HOW TO FISH

Paul: Come on Robert, pay attention!

> 'Angling or float fishing I can only compare to a stick and a string, with a worm at one end and a fool at the other.'
>
> *Samuel Johnson*

BOB

A lot of people say it's impossible to learn to fish from a book. They reckon you can only learn by getting out there and doing it. It's a bit like darts. If you read a book on how to play darts at an elite level, it'd read:

1. Buy darts.
2. Find dartboard.
3. Always hit treble 20 then the bullseye.

This is the complete text from my 2012 book, *Bobby 'Motown' Mortimer's Complete Darts Course: Lessons From the Deart (Dart/Heart).* Now, while those instructions are all technically correct, it's not going to make you a fantastic dartsman (something that was established in subsequent court cases brought by angry readers). For a start, it mentions nothing about all the lager you have to drink to become truly good, or that you'll need a smart nickname, like French Fries, Bobby Carpets or the Puzzler.

Similarly, fishing can be boiled down to this:

1. Buy a rod.
2. Catch the country's biggest carp.
3. Buy a crown and pose for *Angling Times* with the crown on your head and that big trout in your arms.

When I was younger, fishing between the ages of 12 and 15, I'd waste hours listening to lots of rumours from other children about how you can definitely catch a fish.

One of them was tickling a trout. I've never discovered it if it works. People always claim it does, but I've never seen any evidence.

The next rumour was that if it's dark and you shine a torch on the water, the fish will come up and then you grab them. Again, I can't believe this would work, but as a kid, it seemed very convincing. Some older kid would tell you this, nodding his head sagely, telling you with confidence, 'Oh, Bob, man, it's *guaranteed*.'

The next one is that you find a pool and you pour bleach in it, the fish die and float up to the surface and then you pick them out and then eat them.

The next one – and you never get to do this, because you're just a kid – is that you throw a stick of dynamite in. All the fish then float up to the surface. Funnily enough, this one is true because I've seen a thing on YouTube where they were blasting a bridge or something. You looked down at the river – BOOM! The blast goes off and slowly, all the fish come floating up. I don't know if they're dead or dazed, but it's effective. But of course getting dynamite has become tricky since Maynard's shut. You can't get dynamite any more.

But there's a bit more to the art of fishing than that.

The first thing I suggest you need if you want to become a fisherman is to be friends with Paul Whitehouse. That's because when you're starting out you need an actual fisherman to teach you the practical side. You'll need to know from someone more experienced if you're doing it correct enough that you stand a chance of catching a fish.

If I'm not sure what I'm doing, it erodes my confidence entirely. It's impossible to stand quietly on a river for six hours when you're full of doubt about whether you're doing it right.

There's no shame in finding the mechanics of angling a bit mystifying. I took my eldest son out fishing once, about 12 years ago, and it was a *disaster*. I thought I knew the basics – you know, how to put a float on, how to attach a couple of weights, and then slap on a reel. But it turned out I didn't know how to tie the float, I couldn't work out where the weights attached, and I didn't know how to use those reels with the little arm on them. I just didn't know how to do it.

When we're not doing fishing for the show, Paul has been a superb teacher. I might be wracked with fear – 'Am I doing the right thing? Is the hook right? Am I lying on the bottom?' But having Paul come over, give it the once-over and say, 'No, that's the right set-up' – just hearing those words, and I'm made up.

I'm so lucky to be going with Paul because I don't think I'd like to be stuck in a river with a companion who isn't so nice, calm and amusing. When he's told me I'm doing everything theoretically correct, and I believe there's fish in the water where the end of my line is, I know I'm not completely wasting my time and I'm away for hours. Paul often says to me, 'You've got as good a chance as me.'

I would love to get good enough – and I'm still not – so that I could phone someone up and say, 'Do you fancy coming fishing?' and when they say, 'I don't fish,' I can say, 'Well, I can show you the basics.' I'd *love* to.

As I've got older, I'm not in that game any more where you'd call up a mate, say Mark Lamarr: 'Hey, Mark, are you going to this thing on Saturday? Are you doing that thing on Tuesday? Shall we go to the Groucho and get pissed?' I'm not in that world any more. But if I had the skill, I could say, 'Mark, you don't fancy coming fishing, do you?' It'd be a nice thing to do. I really wish I could. But I'm not there yet though.

I'd happily go fishing with most people, but the person I'd love to take most is Matt Berry. He's great, Matt: a reflective, kind, interested character who'd deeply appreciate fishing. But I wouldn't want it to be a disaster, like it was with my son. I wouldn't want it to be, 'I'm so sorry, I don't know how to do it.' Everyone coming back, whispering to each other, going, 'Don't ever go fishing with Bob Mortimer. It's fucking awful.'

I'd suggest the novice starts with the fishing that brings me the most happiness. As I've already mentioned, it's the Huckleberry Finn sort of fishing I love. For this, you just need a rod with a reel, a float with a hook, and a worm hooked on the end of it. The key skill you must possess is the ability to get your line in the water. There's only one guarantee in fishing, and that's if your hook's not in the water, you won't catch a fish.

You don't need tons of equipment when you're getting started. Even after two series of *Mortimer & Whitehouse: Gone Fishing*, I've not amassed that much kit.

My most important tool is my old blue rod. It's a little plastic blue rod I bought at Argos for that occasion when I tried to see if my son would like fishing, so I got it about 12 years ago and it was only used once. I took it to the first episode we ever did of *Gone Fishing*, as a sort of joke – 'Look at this rubbish rod I've got' – but when I sat down by the bank and there was all this palaver of 'You've got to choose this, choose that', I thought, 'No, this little blue rod is a perfectly good fishing rod.' I suddenly felt very defensive towards it.

I've become a little bit obsessed with it, like it's my first car or something. So it's my fate to have this shitty little blue rod and I'll stick with it. It's not that I'm tight because I've got a nice fly rod – because I can't fly fish with my blue rod. Paul occasionally likes to go fly fishing, so I tag along with him. Paul pointed me to a good starter rod, and the very first thing I did with it was practise in my garden, but my cats kept up a torrent of interference. They thought it was a fabulous game.

I think the fly-fishing thing is a little bit like riding a bike, because I've not cracked it yet. They all say it's just practice, and I'm at the point where one in probably every 12 to 15 casts is of any worth. I think I might be in the area of suddenly I'll give it a cast, and everyone will joyfully exclaim, 'Oh yes, that's it! Bob's cracked it!' I hope that day will come.

The other essential piece of fishing kit is a hat. I go to Palm Springs every year and they've got a hat shop there. It's got a tourist vibe about it – the sort of place you can get holiday hats from, novelty cowboy hats, that sort of thing. But I was on one of my days out, so I went inside and it turned out to be a terrific hat shop. A *terrific* hat shop. We had the new series of *Mortimer & Whitehouse: Gone Fishing* coming up, so I picked up six new hats. One for every episode.

It's partially when you're getting older and your hair is diminishing – and I don't mind it, I don't buy hats to hide what nature has abandoned, and I'm not at the edge of some sort of breakdown – but sometimes when you watch yourself back on a show, and you're being filmed from all different angles, the TV screen suddenly fills with the back of your own head and you think, 'Oh my days …' So, I'll be honest, it was my own selfish motive to occasionally have a hat just so my balding head wasn't quite so apparent. And as long as the series continues, an annual visit to the Palm Springs hat shop will be one of my traditions.

When I was younger, I was a lot more sensitive about my bald spot and for TV appearances I would apply hair-thickening spray to cover my embarrassment. I became highly skilled at this deceit.

These skills served me well during the filming of an episode of *Mortimer & Whitehouse: Gone Fishing*. Paul had mentioned a few times during filming that his bald spot could potentially be dominating the shot at times, especially when the cameras were behind us, which they often are as we are generally facing the river. So on this particular trip I brought a can of cheat spray with me and gave him a bald spot transformation. He was very reluctant and sceptical, but when he saw the results he was in shock: 'I look like Tony Blair when Tony Blair is viewed from behind Tony Blair.' He was quite chuffed, I think. We went to the pub that night with Paul's new hair intact and I sensed a real spring in his step. We also ordered a steak and kidney pie, which further elevated his mood. No one mistook him for Tony Blair though.

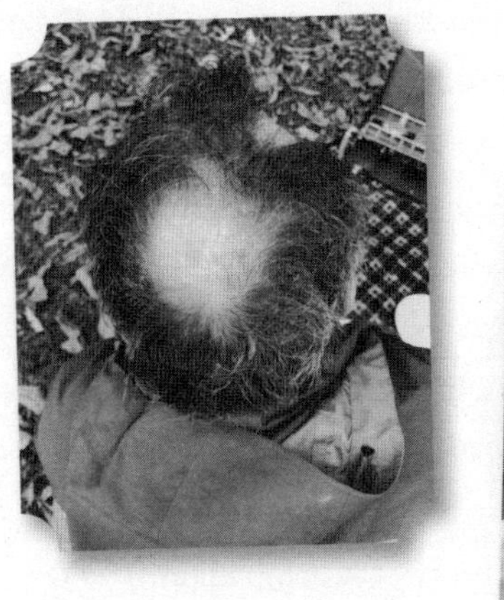

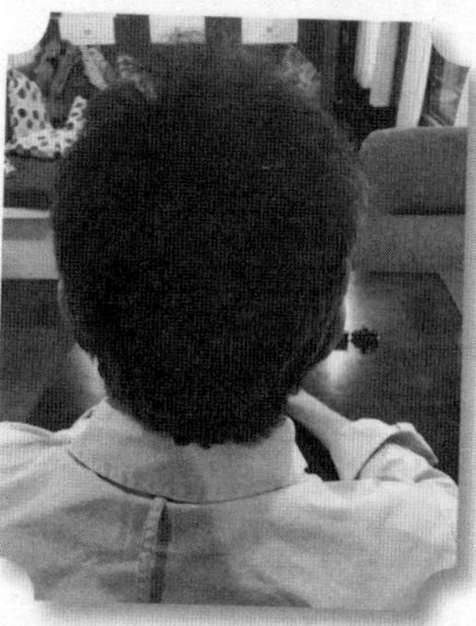

Bob: Paul's cheat spray results before and after.

So you've got your rod and your hat. If you've chosen whereabouts you want to fish, then find out if there's an angling club nearby. Gently approach a fellow angler down at the riverbank and see if he can give you some pointers. By and large, fishermen love to have a chat – usually because half the time the fishing's not the greatest, so a brief moment of human interaction can be a bright spot in their day.

It might sound obvious, but if you find fishermen, there's usually fish. They wouldn't be there otherwise. If the fishermen are out, then you'll also know the conditions are decent for catching, whether that's the weather or the time of day. But approach the fisherfolk with a note of caution. I've met a lot of fishermen through the show and they're often extremely competitive with each other. I've watched them – one will say, 'So, I'm using a nylon for that,' and the other one will say, 'Oh, you use nylon for that? Right. Only I use braid.' They undermine each other constantly. 'Oh, you're using *that* to fish, are you? No, I mean, whatever suits you … only I've heard it's not that good.' I sometimes think the one thing that's nice about the show is to remind people, for goodness' sake, that the only motivation for a fishing trip should be to have a pleasant day.

Anytime we do something a little bit specialist on the show – like carp fishing, or sea fishing, or our first trout fishing trip – we might get an expert along who knows that river and knows the sort of tricks that those fish are more likely to fall for. You never waste time talking to an expert.

Once you've settled quietly and are nestled in a wooded little thicket by the riverside, hiding from the fish's excellent vision, then you try to think like the fish. Where would you go if you were a particular fish? Time teaches you that chub are often quite near trees, pike are near the reeds, that sort of thing. Think like the fish: what would you eat if you only ate things that fell into the river at this time of year? Study the terrain. If you were a tench, what would attract you, what would startle you, what do you truly desire? You are locked in a battle of wits and wills with your quarry. Start to get inside the fish's head. Know him better than he knows himself. Become one with the fish. Seduce his wife. Dance for him at a family gathering. Whatever it takes.

Basically, you're hoping to increase the odds that when some inquisitive or greedy fish makes a mistake, your hook will be there when it does.

It can be a bit sedentary, but the moment when you strike and there's something on the line and you feel that life force running through the line and down the rod into your hands … well, you can't put it into words. No matter how big or small it is, it just makes your heart skip a beat.

To begin your coarse fishing education, it gives me great pleasure to put you in the capable hands of my personal fishing instructor, Paul Whitehouse.

PAUL

Before we crack on here, I'd like to issue a disclaimer: there have been a million and 73 books written about how to catch fish and at least a million of them will go into greater detail than I could. But a certain diminutive Northerner with a head like the moon once tried to take his sons fishing and realised he was quite incapable of even setting up a rod, reel and float tackle. So, in the very remote chance that there is anyone as 'challenged' as him out there, this is an attempt to cover the basics of bait or coarse fishing.

I apologise in advance to all those experts for whom this will seem basic to say the least and any readers who are more interested in one of us dropping dead (I've placed my bet) than reading about waggler floats and shotting patterns. But Bob sometimes demonstrates an iron will and even I bend to it occasionally. In fact, I almost told him to 'do one' because there are endless ways to fish, in a bewildering array of waters, for a large number of species, that feed in different ways (some don't feed at all in freshwater) from one day to the next – never mind in different conditions during the season. I mean it's not brain surgery … no, it's more complicated than that. And rocket science. But Bob's a mate and a fairly nice bloke really, so here it goes … (Please see page 100 for Izaak Walton's disclaimer and forgive me).

FLOAT FISHING

Float fishing is, unsurprisingly, the art of fishing using a float. Bob and I love float fishing, because it evokes our childhood worlds when we first discovered fishing. It feels like the most traditional method of fishing that we still employ today.

Paul: A collection of floats.

The float is the indicator that shows you when a fish has taken your bait. It sits on the water surface until a fish takes your bait. That's the moment a fisherman's heart leaps into his mouth. *It's there! It's there! It's come!*

Floats come in a bewildering variety of shapes and sizes but there are two or three types that will see you through most situations when starting out. When fishing still or slow-moving water you would usually go for a 'waggler' – this is shaped a bit like a pencil.

The waggler gives less resistance when a fish takes your bait, as it has a small surface area. This means a fish is less likely to be spooked when it gives the bait a little exploratory nibble, as it doesn't encounter any resistance from your float. In addition to the waggler, for very sheltered spots and closed-in work you could opt for the more delicate quill float, but the principle is the same.

If we're going to fish still or slow-moving water with a waggler, you fix the float by the bottom end only. Most floats have a ring on the bottom so you put the line through that split ring on the bottom end only, and lock it in place with two split-shot weights.

You can also use a little silicon tube just above the ring, so the line goes through the silicon tube, which then slips very tightly onto the bottom float. You don't need to lock it in place with the split-shot – they'll be added further down the line to get the bait down to where you need it. That's usually, but not always, on or near the bottom. If you're trotting, which is letting your line run with the flow of the river, you'd use a different type of float, certainly in faster-flowing water.

The profile of the float would still be basically a pencil shape but there is a wider body or bulge near the brightly coloured tip. This helps the trotting float 'sit' nicely in the water.

When trotting, you need to attach the float at the top and the bottom by using silicon tube or similar. This allows you to control the line on the surface and make adjustments so that the float and your bait can run freely through the swim. Please note that this is a very basic set-up and description of float fishing. There are many different rigs and techniques, but for the beginner these are the two different set-ups for still and running water.

Generally to start with, you will be using single-strand nylon line or monofilament. The breaking strain of the line you use depends on the species you're fishing for or how many snags and obstacles there are in the water you're fishing.

For smaller species like roach and dace you would use 2- or 3lb breaking strain line, maybe even finer for the hook length, increasing in strength for species such as tench and chub to around 5lb and obviously going up again for the big boys like barbel and carp and for the scary ones – pike – anything around 10- to 12lb breaking strain or higher.

A lot of anglers use braided lines these days too but these are often for specialist rigs and hair rigs, knotless knots and set-ups that would cause Bob to have a full mental breakdown, so we'll omit them here.

Line comes with its diameter on the spool or packet – 0.38mm is the one I've got in front of me – and that's important to fishermen, because they're looking for a strong but extremely slender line, in the hope that the line won't look obvious to any passing fish.

Then you have to attach your hook to the end of the line. There are millions of knots you can use to do this. A lot of people – Bob, for one – use the improved clinch knot.

I know a lot of fishermen swear by it, but I would never use it. There's a knot called the Grinner, or water-knot, and it's much better than that. It's a brilliant, versatile knot with great strength for tying your hook or attaching two lengths of line together. A lot of people will proudly tell you this means you only have to learn one knot for attaching your hook. *Except you don't!* You have to learn more: it's the law.

The type of hook you want often depends on the type of fish you're setting out to catch, but they're all sold clearly marked and often in huge packets or plastic tubs (often including little weights and floats too) for next to nothing – although it's quite easy to spend a fortune on next to nothing where fishing is concerned; I've done it many times.

Most anglers these days use barbless hooks. You can buy barbed hooks – they have a better chance of sticking into a fish's mouth because they penetrate more deeply, meaning the angler will lose less fish once they get taken. But barbed hooks also do considerably more damage when they're removed, especially by inexperienced fishermen. With anglers being fanatical about caring for their fish and wanting them to be returned to the water unharmed, it's barbless hooks all the way.

The most common hook is J-shaped, but a lot of people now swear by circle hooks – these are like J-shaped hooks, but the hook point is slightly offset. This is supposed to help hook fish effectively – as they swim away after taking the bait.

In float fishing, you use two main types of reel: a fixed spool and a centre pin. The centre pin is the old traditional reel, what a non-fisher expects a reel to look like. It's a circle, there's a spool and the line simply winds round it and onto the spool.

With the fixed spool the line is caught around the bail and then wound evenly around the spool. The advantage of the fixed spool is that you can cast much further with it. It also has what's called a slipping clutch, which means if and when you hook a fish, you can adjust how much line the fish can take without the line

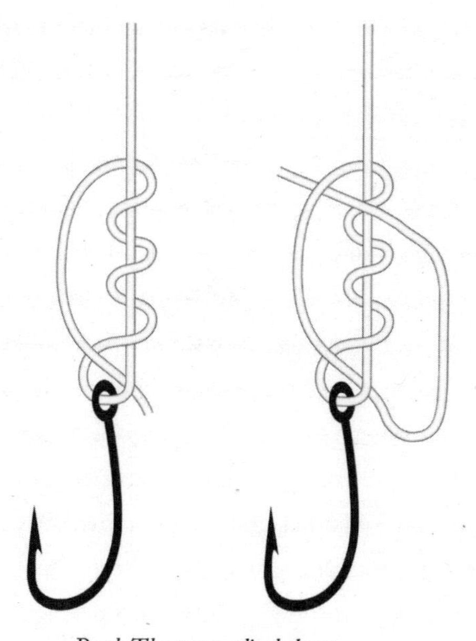

Paul: The trusty clinch knot.
But not as trusty as the grinner.

snapping; by adjusting this clutch, it lets the spool run freely even if you're still winding. This is essential if you're Bob Mortimer and you're winding furiously, despite being endlessly told not to, shouted at and occasionally kicked and punched, while a fish is heading in the opposite direction.

The centre pin gives you more direct control but you have to adjust the resistance that you're going to give the fish with your hand.

So they're the two reels. Either will do for either situation, but for real distance when you're casting, you need a fixed spool. For sensitivity and line and float management in a river, a centre pin gives you more control.

In terms of rods, when you're float fishing in a river or a lake, really anything from 11 to 13 feet is about the norm. In a way, the longer the rod, the more line control it allows – especially if you're trotting; you can keep your line off bankside obstructions or around weeds and reeds. The length of rod is not so important in a lake as your bait is usually fixed in one position.

There are various test curves, or strengths, of rod. So if you were fishing for something like roach or dace or grayling – not a particularly big fish – you can use a fairly light rod. If you're fishing for carp or barbel, chub or pike, then obviously you use a much stronger rod with a bigger test curve. The test curve is the amount of weight required to bend the tip at a right angle to the butt (Ooh, Matron – it's the handle of the rod, effectively). But you do it by judgement as much as anything.

Then you need to plumb the depths. Not the moral depths that Bob and I have trawled in the vain hope of a gag over the years, but the depth of the swim you're fishing. You do if you're float fishing. If you're ledgering, your bait is already going to be on the bed of the lake or river. But even with a float, you're almost invariably fishing on or near the bottom, because that's where most fish lie, so you want to make sure your bait gets as close as it can to the riverbed. Trout are different, because they'll rise to a surface fly. Chub will do that and carp will also take a surface bait, even though they're predominantly bottom feeders.

Predators like pike and perch will actively seek out their prey in mid-water, but the majority of the time, you're pretty well fishing on or near the bottom.

Plumbing can be tricky. That's why a good one is worth its weight in gold – but that's enough lame plumbing humour. Plumbing the depth at which to set your float so your bait trundles enticingly along is tricky because riverbeds vary: One section of riverbed in your swim might be a nice uniform two metres deep but then suddenly shelve up and your float can get caught on the bottom.

You start by putting the plummet onto your hook – plummets are little lead weights with a cork base, so you can just push the hook in. The float has a capacity of weight that will sink it, so you effectively overweight the float. The plummet sinks the float. You cast out, the float sinks and disappears. So you move the float up the line – you cast, it still sinks. So you move the float up another foot and eventually, it emerges.

You want the float in a position where the lower part is submerged and the tip is visible. If you overdo it, then the float will just lie flat until the current takes it. It's trial and error until you get that depth about right.

Bait your hook, hurl in some offerings of what you're using as bait (this is called loosefeeding) and off you go. Keep loose-feeding little and often to draw fish into your swim and keep their attention.

LEDGERING

Ledgering is fishing on the bottom with a fixed weight.

It's another, possibly even more productive, way of presenting your bait to the fish, than float fishing. You use a simple weight to anchor your bait or a swimfeeder, in which you can pack some of your particles of bait, or groundbait, to give a suggestion of what's on your hook, or even exactly what's on your hook.

For ledgering, you'd use a rod round about 10 to 11 feet and you can use that in rivers or lakes. If you're carp fishing, it needs to be a strong rod, but you have all manner of bite indicators, so the sensitivity of the rod isn't an issue. As soon as you can get a run, you'll get a beep sounding from the electronic bite indicator when the circuit is broken or when the circuit is made – I don't know which way round it is. I'm not an electrician, am I?, but either way, the alarm goes off.

In certain other situations, you would use what's called a quiver tip, which can either be built in or attaches to the end of the rod, and this is a more sensitive indicator. That's what you'll be looking at to see whether a fish has taken the bait. You could also use a swing tip, which is perhaps a more sensitive method than a quiver tip, but better suited to still waters. It's just an inert dangly bit that hangs at right angles and twitches up when a fish takes.

And then there's freelining, which is self-explanatory. You might fish very carefully in the margins for a fish on the bottom and you'd use very little weight, if any at all, so the fish takes the bait and doesn't feel any resistance.

You have to watch the line, but that's tricky – you've got no weight to cast other than the bait, and if the wind's blowing or anything like that, you'll be blown all over the place. When fishing like this, stealth is important, so keep clodhoppers and lumpen idiots (like a certain bloke from Middlesbrough) well away from the fishery. In fact, don't tell him you're going fishing at all. Certainly don't give him your contact details.

The high-end, hi-tech carp fishers go in completely the opposite direction. What they do is use a very heavy weight to anchor the bait. The bait is not far away – it'll hopefully be in the middle of a load of loose feed that they've bombarded the lake or river with. Boilies, which are high-protein manufactured baits, and sweetcorn are popular baits for carp, and the carp angler will use all manner of flavourings and inducements to make them even more tempting. They even use fake sweetcorn, which allows them to cast miles, and the greedy old carp will happily take even that. Mind you, a lot of carp are extremely wary and suspicious of all anglers' baits and many silent men spend days on end around mysterious ponds and lakes catching absolutely nothing.

Carp fishermen will almost always bait or 'spod' the area where they're fishing. A spod is a plastic rocket-shaped tool full of bait that is cast to a particular area, so that when they cast the hook bait, it will be hiding amongst all this additional groundbait.

They often use a big fixed lead, and they usually attach the bait by means of what's called a hair rig. With a hair rig, the hook isn't in the bait: instead, it's hanging on a tiny bit of very fine line next to it. The carp picks it up, swims off thinking, 'Oh, this is all right,' suddenly feels the weight, bolts and that engages the hook. It seems like hi-tech stuff but it's quite universal in a lot of ledgering now, but if you're using a small particle bait like a maggot, then you wouldn't necessarily use a hair rig. If ever you find yourself in a pub and a funny little rotund Northern bloke engages you in conversation about fishing, bewilder him with terms like 'helicopter rigs', 'zig rigs' and 'simple bolt-hair pop-ups'. A lot of carp rigs can be purchased ready-made now. You just need to attach your pre-tied hair-rig hook length to the swivel at your weight if you're ledgering or to your mainline if you're float fishing. It'll save you learning the knotless knot!

Carp will also take surface baits. Floating crust was an old favourite way to catch them and it's very exciting to watch a big carp slurp down your bait from the surface. These days particle baits like certain types of dog biscuit and pellets have superseded the old floating crust, but I have fond memories of fishing in this way.

SPINNING

The tackle you use for pike and perch is slightly different. As predators, you can also catch them on spinners, light lures or jigs, which are small, highly coloured little artificial fish that are often articulated so they move about alluringly.

There's a bewildering array of spinners, lures and plugs. The amount of money you can spend is incredible. It's a nice way of fishing – it's roving, you're walking, you're casting into likely spots, trying to make this lure seem alive with careful jiggling of the line … it's very exciting.

You'd use a lightweight spinning rod of about nine or ten feet with a fixed-spool reel – it excels at light bait casting. You certainly wouldn't use a centre pin with a light lure.

If you're fishing for pike, which can be anything up to 40lbs, then you need a much heftier rod – say nine to ten feet if you're spinning or lure fishing. The pike likes to ambush its prey, so it will wait in a place where it can dash out. Like the Usain Bolt of the fish world, it likes to use a sudden burst of speed over a short distance to catch its prey.

You'd use your lure or dead bait, and they like all sorts of dead bait, pike, especially sea fish, like herring, smelt and mackerel, but one of the best baits for a really big girl (all big pike are female) is a small pike and a fairly hefty fixed-spool reel. When fishing for pike, you also need to use a trace on your line because otherwise a pike's teeth will bite through nylon. But there's loads of specialist gear for pike fishing. Anglers – well, most of them – love kit! And tackle companies are very happy to provide tons of it. Most of it is designed to catch anglers rather than fish, but it's hard not to be seduced by some 'new' development that will guarantee you'll catch the next record pike … come on! Many books have been written about pike and pike fishing, not least because they are the 'Jaws' of the freshwater world. Top predator. Dead-eyed. Scary. Fantastic.

There's another technique called drop-shotting that has become very popular in recent years, especially for perch. You use a weight at the end of your line and the lure is suspended at a right angle above the weight. You cast, the weight takes it down to the bottom, and the lure sits at an enticing right angle some distance above it.

CASTING

No one can learn to cast without getting out and casting until you start to get the hang of it. It's not possible otherwise. But in the interests of what you'd expect in a book about fishing, there are two main casts depending on where you're fishing: the overhead cast and the side cast.

THE OVERHEAD CAST

To cast, point the rod at your target and then lift it slowly backwards in one smooth motion so that it's slightly behind you. Imagine the rod is the hour hand on a clock face and stop around one or two o'clock. Then bring the rod forward, stopping briefly at about 11 o'clock, releasing the line from your index finger and slowly following through with the rod so that it ends up at nine o'clock and you're ready to fish. If you're fishing a still water, click the bail arm over to stop the line running off the spool once you've cast and fish away. If you're fishing a running river and trotting your float downstream, leave the bail arm open and control the line with your finger to allow the float to trot with the current and your bait to swim down naturally. The same basic principle in casting with a fixed spool applies whether you're float fishing, ledgering or lure fishing.

With a centre pin, the beginner draws off as much line as possible and flicks the float out. Unless you have mastered the Wallis cast, an extremely difficult technical cast that requires much practice and months of overruns and tangles of bird's nest proportions, you're better off fishing with a fixed spool to begin with.

Casting doesn't require brute force or getting the weight of your body behind it. You're only ever using your arm as a pivot. Imagine you're a robot: you want a smooth, controlled and swift motion.

Paul: Be quiet, Bob, you're scaring off the fish!

THE SIDE CAST

Not everyone is going to be fishing on a large river where you can make long, arcing overhead casts so most beginners will probably find the side cast the easiest to start with.

It's exactly the same principle as the overhead cast, but the rod starts, as you might imagine, from the side and you use a similar action that you'd use if you were skipping a stone across water. Again, the rod should do most of the work.

The side cast is great if you want to get under overhanging trees – you just skim it in – and because you're coming in at a lower angle, the bait or lure makes less of a splash when it hits the water. It's also effective on windy days, which can play havoc with an overhead cast. But like all casting, it takes practice before you'll be accurate enough to get the lure to land exactly where you want.

Nothing will help you as much as getting bankside tuition with casting, though these days there are many brilliant online tutorials that can really assist the beginner and experienced fisher alike.

So now all you need is to get out and try it. What's that? You just caught a perch? On your very first go? You jammy sod. Well done!

SOME OTHER CONSIDERATIONS WHEN GOING FISHING

BAIT

Like religious people, different fish like different morsels to eat, and some species of fish won't eat things that other fish absolutely love. Unless you're fly or lure fishing there's no avoiding it; at some point you'll have to get messy with bait.

As you can imagine there are many baits: some inert, some wriggly, some smelly and some you prefer to your sandwiches. The big five, though, are bread, worms, sweetcorn, luncheon meat and, more recently, boilies. And pellets. Okay, so there are six main baits. Actually, there are seven because at number one in the poptastic bait charts, great mates, is the humble, larval stage of the bluebottle and other flies, the Mighty Maggot. That bloke out of Goldie Lookin Chain knew what he was doing when he called himself Maggot.

The maggot rules supreme and that includes the caster, which is the chrysalis stage of the maggot before it emerges as a fly. The caster is a great bait in its own right and has led to the downfall of many a specimen roach, chub and carp. Mix it with hemp and you have a heady cocktail that has lured many a barbel too. But the maggot or gentle has a special place in the angler's heart and a not very special place in anyone else's heart. The hardcore fisher thinks nothing of eating their packed lunch having been plunging their hands into a bucketful of our wriggly friends all morning.

And surprisingly, there is very little incidence of E. coli, MRSA or C. difficile after such behaviour. Almost every fish that swims in freshwater will take a maggot in any kind of fishery, still or moving, and that's why it reigns supreme, but there are a couple of pretenders to the throne that have muscled their way onto the scene in the last few decades, namely the boilie and various types of pellet, halibut being possibly the most popular.

The boilie – a manufactured high-protein, flavoured ball – has become ubiquitous due to the explosion in popularity of carp fishing. One of the excellent properties of a boilie is that it can't be nibbled away at by smaller fish, so you know that any bite when it comes is likely to be your target fish, mainly carp. Although barbel fishers and tench and bream anglers were not slow to catch on.

Sometimes, though, the boilie can be overused, and the carp become wary of the continual bombardment and use of such baits (some carp fishing is like a military exercise); in which case a quict approach, with a natural bait, such as a worm or a bunch of maggots, can pay dividends. I remember going to fish a carp lake with legendary angler and writer Chris Yates one day and we used a kind of freelined maggot approach. It didn't do us any good, but it was refreshing to see. As Chris put it, quite a lot of carp fishers declared war on the carp a few years back, so it was nice to fish in a simpler, less invasive way. I suppose the trick with all angling is to be versatile. Later that day I caught a carp on sweet-corn: final score Whitehouse 1, Yates 0. Mind you, Chris also took me for a day looking for crucian carp on a beautiful hidden lake somewhere that time forgot and he absolutely battered me. No contest.

Never forget the worm as bait. It has been an excellent and consistent bait for most fish for centuries. Many anglers have spent the night before their fishing trip in their garden, wearing a dressing gown, armed with a torch and pulling up lobworms (the Americans call them 'night crawlers', which makes them sound much more exciting). The best way to bring up big lobworms is to hope for a downpour or create one yourself by watering a lawn. Trying to catch lobworms at night is often more fun than the fishing trip the next day. I remember once being immersed in a giant compost heap on a remote farm in Wales, having become dangerously obsessed and careless in my relentless pursuit of brandlings (a type of worm) for a sea-trout session. How I laughed when I realised I was stuck in a vast pile of cowshit as the crust gave way. Oh, the joys and folly of youth!

Most tackle shops stock artificial substitutes – like soft plastic jelly worms in a million different sparkling colours – and synthetic baits, all of which come coated in mysterious scents that seem to attract fish (while these are an industry secret, many anglers believe they're extracts of fish roe). The synthetic baits often come in little plastic tubs with branding that make them look and sound like big canisters of vape juice – some of the most popular brands are Quantum Radical, Mainline, Heavy Metal and Dynamite Crave.

If your quarry is a predator like the perch or pike the other very popular option is a lure. These are cleverly designed artificial baits which display certain movements, shapes or colours that imitate the prey of the fish. They come in all shapes and sizes, and most fishermen carry a selection that work in all zones of the river, from the surface to the very bottom. Crankbaits, spoons and plugs look like little fish; retrieve them in various different ways to give the impression of a sick or injured fish which will attract your prey. Some of them can be extremely effective.

So choose the appropriate bait, fish it with panache and confidence. Some days you persevere with your first choice and build up your swim with groundbait or loose feed, other days you will ring the changes. Only experience will tell you.

LANDING NET

Once you catch a fish, you need to have a net ready for when you land it. All fish should be landed carefully. Make sure that you have a clear, snag-free area to submerge your net and to guide your quarry towards. Beware the fish's last dash when it sees you and the net, and draw the fish over the net and lift smoothly. Hurrah. Result. Touch. Gertcha!

Technically, this is the only time you'll be an actual fisherman – fishermen use nets, while anglers use rods. You can use the words interchangeably, so tell this to an angler in a pub and you'll either have a great bit of trivia to start a friendship, or he'll disagree with you and the insults could lead to a fight. There's no middle ground.

A Knife

The confident angler needs a knife to cut bait, cut lines and, if necessary, to wield in rural locations if people give you a look that irks you. Just stand up and brandish the knife at them from the opposite riverbank and they'll scurry off immediately. No, actually, don't do that. Don't carry a knife around with you at all – you're not in *West Side Story*. Just take some needle-nosed pliers and some nail clippers; they'll work just as well.

Before You Leave the House

Before you ever step foot at a river with your tackle, you need to have a fishing licence. If you're fishing with a rod for freshwater fish, salmon, trout, sea trout or eels, you have to have one. They come in lots of different durations (from one day to 12-month versions) and cost between £3.75 and £27. If you're under 16, they're free. If you get caught without one and prosecuted, the maximum fine is £2,500. Anglers hate illegal fishers. They won't hesitate to dob you in.

Check the calendar. The coarse fishing season is closed between 15 March and 15 June. This is the time when the coarse fish spawn. It applies to all UK rivers and streams, but not every lake or reservoir – check before you go. By the time the season starts again, the fish have had their last spawning season and they've been recovering, building up for the next one in the spring, so they're full of randy, frisky energy.

Check the weather. It's true, you can catch fish all year round in all weathers, but when the weather goes too far one way or the other, it makes it harder. Too hot, too cold, too wet, too dry – all these things make the fish shrink away deeper into their underwater realm and are less likely to maintain their normal pattern of feeding. So get the weather right and you'll be more likely to make a catch.

Then check the time. Most freshwater fish are crepuscular feeders, which means they're most active at dawn and at dusk. Again, you can catch a fish at any time of day, but sunrise and sunset are the times when your odds of them being out and on the prowl for food are highest.

Put your hat on as you walk out of your front door. It marks the moment you transform from an ordinary, dreary, dull working man into a dashing, noble, carefree fisherman. Puff your chest out, take big strides and shout a cheery hello at the sun. You may also click your heels at this point, should you so wish.

Tell someone where you've gone. If you're getting on a bit, they'll be worried that you've wandered off all confused in your dressing gown and slippers and the police will be out searching fields and woodland with the cadaver dogs before you know it.

DOWN AT THE RIVER

First off, you need to know where to go, where there's fish and access. The easiest way to do this is to look online or find a local tackle shop, pop in, and ask where the owner would recommend you start. Maybe sweeten him up by buying a pint of maggots first. Whatever you do, don't book a week on the Tay at this stage. It'd be like buying a pair of Middlesbrough FC slippers and then waiting for the call-up against Barcelona.

Once you get to the place you've been told about, you need to find a suitable place to fish from. Approach the bank quietly and calmly and wear clothes that blend into the bank – mossy greens, greys, dark browns. Basically, you want to look like a woodland elf (pointy ears are optional). Find a shady place where there's a bit of foliage that you can blend in with. Fish have great eyesight and good hearing, and they're easily spooked. No one's quite sure how fish's vision or hearing works, but they do seem to be living in a state of permanent anxiousness so anything you can do not to frighten them works in your favour. Fish definitely seem more sensitive to bankside vibrations, heavy footfalls, and northerners mucking about rather than the sweet voice of a London-Welsh angling/comic colossus. Remember to check the ground before you sit down for discarded contraceptives, empty cans of Tyskie Polish beer and dog dirt.

Please observe angling etiquette and consider the other fisherpeople. If you are fishing around others, remember to keep your distance. First off, it's bad form to sit near people who are already there, because you're essentially squatting on the place they've found and splitting the chances of them getting a catch. Don't cast close to where they are – your lines might get tangled and in my experience, fishermen who are unknown to you can get quite angry and frustrated when you interfere with their day out.

PREPARING YOUR FISHING SPOT

To pick your spot, you need to think like the fish. If you were in a river, you wouldn't hang around in the middle where the full force of the current is rushing at you, because you'd have to swim really hard just to stay still. You'd be in one of the spots of still water by the bank, under overhanging tree roots, in a little eddy or relaxing in a small oxbow lake. Imagine you were hanging about Oxford Street at rush hour during the height of summer. You wouldn't be standing in the middle of the busy road: you'd be in the cool shade of the big Sports Direct or the shop opposite the big Primark that sells the Kim Jong-un masks. Where would you go when it's cold, or really hot, or not really either, but you're a fish and you can't leave the river so you've got to go somewhere? They have to be lurking *somewhere*. Think like the fish.

To encourage the fish to take your bait, you might consider baiting a spot. This means starting off by chucking a load of your bait into the river to get the fish used to it, and hopefully feeding on it merrily before you slip your hooked bait into the waters. *They won't suspect a thing.* Some fishermen will bait a swim for days beforehand.

LET'S FISH!

It's important before you start angling that you're aware of the policy of catch and release. This means after you've caught a fish, you release it back into the water as quickly as you can, and it's a policy that the vast majority of anglers follow. Catch and release means the fisherman has his fun, but the fish is allowed to get on with its day and fish stocks aren't depleted. It also gives you the thrill of seeing the fish that's brought you so much pleasure go happily swimming away afterwards, and you can sit back and congratulate yourself on your benevolence and kindness. Avoid handling the fish too much: wet your hands, get the hook cleanly out of its mouth using your pliers (or another hook-removing tool if you have one), don't squeeze it, don't keep it out of the water too long and when you put it back, hold it in the river facing upstream for a bit to let it get its breath back. Marvel as it suddenly swims off with a flick of the tail, to tell its fish mates that it managed to outsmart you and get away because it's so clever.

ONCE YOUR BAIT IS IN

Sit back and wait quietly. Listen to the birds. Watch some beetles in the dirt. Think about the dinner you're going to have later. This is the best moment of fishing when you've done all you can and now it's up to the fish. But concentrate and fish with confidence. Believe! Yes, believe.

Keep your eyes on the float – and be alert to the way it's acting. Just because it jerks, it doesn't mean you've got a fish. Sometimes the current will make it suddenly bob up and down, or a fish will give your bait an inquisitive fleeting nudge, so don't be too hasty to strike. Keep your eyes on it. How's that float behaving? If it goes right under and stays under, bingo. If it starts moving against the water current, hello. If you're not sure – strike anyway. Similarly, with a quiver tip, if the tip pulls round smartly, strike, but if there is a series of knocks, wait. Should the knocks continue, it might be worth striking; you never know, you might hook the bottom or the fish of a lifetime.

If you've been sat watching your rod for a quarter of an hour and there's been no action whatsoever, try casting somewhere else. Or re-bait and cast to the same place, and allow the glorious process of fishing to begin again.

Taking a chair along with you is a real bonus. Paul and I fished for perch from a chair, and it made such a difference. Maybe it's an age thing, but you're not just there to fish – you're there to enjoy the day, and people can easily forget that. And you're there for six or seven hours, not two, so why not have a bit of comfort? Even a train seat isn't that great after three hours, so imagine how much worse sitting on the ground is. It just makes the day work, being comfortable – although it's a special treat. Sometimes you have to merge into the foliage when you're fishing, so you can't just have chairs set up all over the place.

CATCHING THAT FISH

Once your float dips purposefully under or slides away, or if you're ledgering, your rod tip pulls round, you have a bite. A fish has taken the bait. Result? Well, not yet. If you're fishing for smaller, more nimble fish like roach, or dace, bites can be quick, a sharp dip of your float and you need to respond sharpish or the bait can be ejected. Crucian carp 'takes' often register as small movements on the float. Bigger fish such as chub, tench, carp and especially the barbell leave you in no doubt that they've taken your bait (sometimes your rod lurches round dramatically with the latter). The angler then needs to set the hook, or 'strike' as it is usually called. But it is not a jerk in the opposite direction or a vigorous movement, it's usually a firm but gentle lifting of the rod into the fish. There are exceptions to this: pike on a dead bait need to be struck relatively firmly; sea-fish at depth need a firm raise of the rod, but with certain carp rigs, like a bolt-hair, you don't strike at all. And if a gillie sees you strike when a salmon takes your fly, he's likely to turn the air blue at your pitiful response (and demand independence for Scotland!). Do nothing when a salmon takes your fly – leave it, I said, leave it! Now gently lift the rod ... what? He's gone? You should have lifted earlier. The gillie is now cursing you in Gaelic and laughing inside.

But let's go back to ledgering for chub, for example. A key part of the strike is timing. The moment you feel the line pull, it's a battle not to strike immediately. But sometimes, the first knock you've had is tentative, with the fish not fully convinced, and before the bite has developed fully, you'll whip the bait right out of its mouth, you fool!* This is part and parcel of fishing – as they say, you've not caught a fish until it's on the bank. But leave it too late, and the fish can reject it altogether or take the

* I'm looking at you, Bob.

bait deeply. If the latter occurs, there are simple tools to use to disengage a barbless hook without causing any damage to the fish. To be honest, nobody can say exactly when you should strike; it's a question of practice and experience, and varies from method to method and your quarry. Sorry to be of no help at all here, but that's life sometimes, kidda – you got to figure things out for yourself.

Once you've hooked a fish, you have to land it, obviously. With smaller fish you do this by simply lifting your rod and reeling in. If, like Bob, you try to employ exactly the same tactics with bigger fish, then you'll more than likely end up with nothing. Mind you, he also screams, shouts and jumps up and down like a maniac, which is entertaining but not helpful when you're playing a double-figure barbel that could easily batter us in a tear up in a pub car park. Let the fish run initially, but maintain tension and keep the rod tip up to absorb any sudden bursts of speed, head-shakes or changes in direction. However, if a fish is intent on heading for a nearby snag, you have to stop it: turn the rod through 90 degrees so the pull on the fish is coming from the side. This will cause the fish to veer away from dangerous weeds or tree roots. Unsurprisingly, this is called 'side strain' and is a good tactic to use in those circumstances. Occasionally you might want to stop a powerful fish from getting stuck in weeds. In which case you have a choice: let the line go slack, which sometimes causes the fish to emerge after a while, or go in! Obviously don't put yourself in danger though. Fish are great at transferring hooks from themselves into tree roots.

The key thing is to keep constant pressure on the line. This means the fish is battling against an unyielding, constant force, and it will start to get tired. As the fish becomes knackered, you will be able to slowly draw it in. This can take a while, but as it's going on, it helps if you try and direct the fish into more open water.

When the fish is beginning to tire, she or he will start to show on the surface and wallow (there's a word). Your landing net is at the ready, isn't it, Bob? In a convenient place with no obvious snags and with the net portion already under the water? Now you need to be careful since the fish will make a dash to escape when he sees you, the net or an excitable portly figure jumping around on the bank. Slowly draw the fish over the submerged net and lift. Voilà!

WHAT TO DO WHEN YOU'RE STANDING AROUND WITH A LIVE FISH

Once you've brought the net containing a fish to the edge of the river, you're going to want to see what you've got and then you're going to prepare that fish to return home.

Fish love water more than they love hot, dry air. That's a well-known fact, so it's important to put them back in the water as swiftly as you can. Every second counts (if a fish caught you, you'd want to be put back in as quickly as possible). The longer a fish is out of the water, the less chance it has of surviving, so it's important to return them as quickly as you can back into their murky underworld kingdom in which no man can dwell. Most

Paul: What a clonker! A cracking barbel that I caught with John Bailey (top) after the cameras had packed up. It's my fish, John …

anglers are pathological about treating fish carefully and returning them as quickly as possible. Many anglers use unhooking mats, which are fish-friendly surfaces designed to keep the fish safe during the process. Carp fishers are especially aware of their quarry's well-being and look after them obsessively, even though carp and tench can survive out of the water a lot longer than other species. I'm fanatical about returning fish almost immediately; sometimes during our series the cameraman rarely has time to get a shot of the fish before I've sent it on its way. It rarely has time to get a selfie with me. Salmon fishers on the River Dee in Scotland (and probably many other rivers) are now encouraged to net salmon and keep them in the net while they unhook them so they never leave the river. Then you hold the fish with its head upstream in the current as it regains its breath before... pow! The salmon kicks away with an incredibly powerful flick of its tail and roars back on its mission to well, you know ... How can I put this? Procreate.

Dry human hands can cause all sorts of problems to fish, from abrasions to rubbing off their protective slime coating, leaving them open to infection. So before you touch them, always dunk your hands in the river or lake and make sure they're wet.

Once the fish is out, what you're going to do is remove the hook from the fish's mouth. This is the very first thing you need to do, and you want to do it as quickly as possible. Barbless hooks come out very easily – you can do it most of the time with your hand, just by backing it out the way it went in – but be careful when you're handling the fish, as one unexpected flick or wriggle and you'll be wearing that hook through the fleshy part of your hand. If you're nervous of doing it by hand, you can use needle-nosed pliers to slip it out.

All being well, you're now getting into angling seriously, so it will be time to pose for your prospective photograph for the cover of *Angling Times* or the *Angler's Mail*; or *Trout*; or *Salmon.* Slip one hand over the tail, the other under the pectoral fin. Don't squeeze the belly, that's no good for the fish. Keep it near the water. Smile. Professional tip: the fish looks bigger if you extend your arms and hold it out to the side with the river behind you, not directly in front of your body. A human being is much bigger than a fish, so your catch will look tiny and pathetic when placed in front of your rippling muscular belly. Or do as I do and get the fish back before anyone can ask for a picture or a closer look.

Time to return that fish back home. It will be absolutely bushed from the massive fight it had when you caught it. Holding the fish as above, lower it into the water and hold it, facing upstream, until it zips off with a flick of its lovely, slime-covered tail. Never just chuck it back into the river. Rule of thumb: don't do anything to a fish that you wouldn't like to see someone doing to your mum.

And that's pretty much it.

Yes, it sounds like a lot, but it will seem less daunting each and every time you go out and have a bit of success. It might be the third time you go that you don't need to look at the diagram of the knot. Within a fortnight, pulling out hooks might be second nature. Within weeks, you might not even think about the stages you need to perform an overhead cast.

Fishing is a beautiful pastime, and like all great pastimes, it won't arrive fully formed the first time you go. After your first trip, you might return tired, confused, bored, listless, damp and fishless. You do have to work a bit to chip away at the sheer strangeness of a day's fishing, but once you do, each and every trip will bring you more joy than the one before. The delight that unfolds with practice and experience is a delight that will never leave you and will only get deeper the longer you do it.

Bob: Here's what you do when you're standing around with a live fish.

CHAPTER 8

MORTIMER & WHITEHOUSE: GONE FISHING

Bob: We'd been going on these fishing trips and had an inkling they might be worth filming.

'If fishing is interfering with your business, give up your business.'

Alfred W. Miller (aka Sparse Grey Hackle)

Bob: I've found a love for fishing and everything that surrounds it, but it took me a long time to grasp quite what it is that people like about our TV show. I know it looks really pretty, so I understand that, and I know occasionally there's a joke or two. But there's not enough jokes so that you could say, 'Oh, you can't miss this, it's *hilarious*.'

Paul: I suppose there's not much on telly that just shows two people who really know each other just having a nice time. The nearest thing to what we're doing is probably Tim and Pru on *Great Canal Journeys,* in which they take narrow boat trips around various locations in the UK and Europe. I like that show and I'm sure that helped trigger our idea.

B: I suppose *Great Canal Journeys* did trigger the idea because we were having lovely days catching fish. We thought we could ask the BBC if they wanted to film it, because that's all Tim and Pru do. We were being genuine; we weren't being arch or anything.

P: A lot of people see *Detectorists* in our show, or *The Trip* – but that wasn't really in our minds. It was more Tim and Pru.

B: Yeah that show did help us. Their show is a day they've spent on the canal. And ours is an accurate record of what happens when the two of us go to a river.

BOB

Even though two comedians going fishing is nothing like *Antiques Roadshow*, that's the feeling I get when I watch the series. That similar early-Sunday-evening feeling. Soporific. That's how it makes me feel.

I know it's regarded as bad form to say you really enjoy something you've made, but I honestly think our *Gone Fishing* shows are just *great.* I think that first series is amazing. I can watch it any time – I could happily go home now and put it straight on. In fact, only this morning I was writing and I put *Gone Fishing* on without the sound. It's quite nice to have it on in the background while I work

I don't mean the show is amazing like *Breaking Bad* or anything, but if I'm not doing anything, it's a lovely 27 minutes. They're so atmospheric and edited so beautifully. I think my favourite

GONE FISHING

A SPOTTER'S GUIDE

BY BOB MORTIMER

Here we have Sergeant Barbel – the Bottom Hugger; a 'lust and cunning fish', as Izaak Walton once said. The barbel can live up to 20 years. It's not a long life, but the possession of those fleshy barbules mean it must be one constantly filled with wonder.

BARBEL

PIKE

Ron Pike's his name, the 'tyrant of fresh water fish' (according to Charles Walker in *The Art of Angling*). Pike have rows of top teeth which are sharp as razors and point backwards, while on the bottom are teeth shaped like needles. It's basically got a knife drawer bin for a mouth. Truly the 'Boss' of the river.

Ah Chester Roach, the Gaudy Merchant. This guy is very similar to the rudd and the dace, and it's easy to confuse them, and quite hard to tell them apart. Even the fish find it hard to work out what's what – roach regularly cross-breed with bream and rudd, the dirty rascals.

Weary McTench – the Bollard of the Lake. The tench is a really hard, tough old fish, which looks rather like a carp that's done a lot of cardio to get back into good shape. They can also live up to 30 years.

Ah yes, the Bream, or as I call it The River Naan. Have you seen one? They're massive! And so very satisfying to catch, too. Paul told me the best way to catch a bream is to fish hard at the bottom. I wonder if The River Naan ever meets Sergeant Barbel down there? What a fight that would be!

The Chubb aka The Clodhopper. The Chubb loves a bit of bread. Seriously, that's all you need. Plus a load of other equipment, obviously. But the Clodhopper is actually the shyest of all fish. You'll be lucky to find more than one in the same spot on the same day!

Here he is, Stripey McBitey aka the predatory Perch. They hang about in the mid-river and eat other fish, so you better watch out if you're a roach or a minnow. Having said that Kingfishers each Perch, so go figure. Ah the circle of life.

Introducing the Rudd or otherwise known as Tin Foil Terrorist. An absolute silver beauty! The Rudd has an upturned mouth to allow it to feed easily on the surface of the water. They can live for over 17 years, which is a long old time in fish years.

Presenting… Raymond Trout, the River Detective! Did you know that trout are capable of looking and focusing out of the corner of each eye simultaneously, meaning they have the ability to see in almost every direction at once? They can vary in colour too! Wow!

All photos © John Bailey

Bob: Here's the photo to prove it! Look at that bass! Well done, Paul!

show was the one where we fished for sea trout in Dorset but I'll happily watch them all. It's a bit mournful that episode, but there's one bit where Paul catches a bass and he's so chuffed. He's so excited, he's shaking and as he turns round, it's affected him so deeply that his face changes, his eyes change, and for a moment he looks like he's a little boy again.

That episode is a lovely marker of what works in the show: us being quite earnest, Paul doing his impressions and the two of us having a right laugh in the restaurant afterwards. It sums up what the spirit of the show is like: enjoyable but ever so slightly dull.

But when we first started, neither of us knew what we were doing. We had no idea what type of a show we'd end up making.

When the first episode went out, I was thinking, 'Oh God, what have we done here?' Say a really bad performance on BBC2 is 500,000 viewers – I thought we'd get 500,000, with people having a look because they like Paul, or they might like me, and then the numbers would drop off as the series went on. In fact, if someone had phoned me up the night before it went out and said, 'The Beeb are thinking of putting this on at two in the morning,' I'd have said, 'Thank God for that.' I was absolutely stunned that anyone watched it. Because I couldn't see it! I could not see it.

When we first had the idea, Paul and I went to see the fella at BBC2. He's called Patrick, and we said, 'Patrick, we've been going on these fishing trips and we really enjoy them, and we've just got an inkling that they might be worth filming.' And Patrick said, 'All right, but what's it about? What's the narrative of the show going to be?'

At that time, TV commissioners were very keen on there being a *reason*. We said, 'Patrick, we've got to be honest. The only thing we can think of is we either catch a fish or don't catch a fish.'

But Paul said we did have a reason: *the jeopardy!* One of us might die on the bankside, and if it was filmed, then that would be a feather in the BBC's cap. They'd be able to use it on the news.

Paul was *very* funny in the meeting, and I think Patrick, just listening to Paul talking, thought, 'You know what? This is fun, sitting here listening to Paul. I wonder if it would be fun if it was a show?' So they took a chance on it. We came out saying, 'Wow! They've commissioned it! And so *what the hell do we do now?*'

Before we started, it was always our intention to talk about our hearts, because that was what inspired us to go fishing in the first place – we wanted to make sure our wobbles with mortality came across. If we hadn't, it might have worked as a show about two blokes going fishing but it wouldn't have had that poignancy, that real *heart*.

So the BBC let us go out there to do the first series, but we weren't quite sure what it was going to be.

We didn't have to attend any production meetings before we started. Paul emailed me where he wanted to fish, I'd get on my laptop and choose an Airbnb (which took around five minutes, all told) and someone would ask me what I wanted to cook, and I'd say, 'Oh, I'll just put some tuna in a frying pan.' And, I might be lying, but I think that was about the extent of the preparation from us, although a lot of people were working very hard on our behalf.

I think the production team were a bit concerned that it might be boring if we were standing on the river the entire time, so we ended up doing some extra bits in case no one enjoyed the fishing. So the crew would say to us, 'Oh, there's a farmer's farm shop nearby – why don't you go and buy your stuff at the farm shop?' We visited a brewery, talked to a very nice vicar – that sort of thing. Very pleasant. But I cannot stress it enough: there was no real plan when we started. So I don't know if the word is serendipity – in fact, I don't think it is – but something a bit special came out of it.

I think one of the reasons people like the show is that it's genuine. Yes, we have the palaver of filming, but at the heart of it, it's just me and Paul. You see exactly what it's like when we're off fishing together. We concentrate for a bit, try and have a laugh, try and treat each other to something. We talk about our hearts and our health and we visit these beautiful places which are right on your doorstep.

For me, the places we visit are one of the best parts of the show. We remind you that all this beauty is just around the corner from you, and these places are so easy to get to; that's easy to forget. And we're saying, 'Look! Don't forget! Look at all this beauty that's here for you to enjoy!'

The most beautiful place we went to was the Derbyshire Wye. As I've said, it was like Hobbit-land, and a place that found a place in my heart almost immediately.

There's a lot of silence in these places and that probably creates its own dynamic. It's almost like being in those peaceful environments gave Paul and me permission to say something serious. It doesn't have to be current affairs – you can say something about your family, or about your health, or whatever. You end up having very different conversations than you'd normally have.

I can't think of anywhere else where these types of chats would spring up. It's a totally different conversation to the pub conversation, or the conversation that you're having watching telly, or the conversation you get when you're in meetings. I can't imagine going to a meeting and sitting there being silent for an hour. Or even the pub. I mean, you do sometimes see couples going very quiet together, don't you, but I can't picture myself doing that. But when you're out fishing, suddenly you're in that unique place, and you haven't spoken to the person who's just there, who's a very good friend, for something like two hours.

You never know what conversations you're going to have out of that quietness. It's really quite stripped bare in that environment. The weather's right on you, there's no one there – but you find a way to talk to each other.

There isn't a script, but we came up with some comedy routines which really tickled me. You'd think the fishing would be an escape, but we do have a responsibility to be a bit funny when the cameras are on. But it's not a big responsibility though. It's nothing to worry about, because all we are doing is just making each other laugh.

It's like the writing experience when I'm writing with Jim – it's so much fun being as funny as you can, but just for each other. There's no one else you're trying to impress. That's how most writers come up with good stuff, I suppose – they're not being judged, the pressure's off and you're just amusing yourselves.

These days, whenever I see a group of young comedians together and they're all trying to be funnier than each other (I'm not knocking them, it's something comedians have always done and will always do), I do find myself thinking, 'Whoa! That was me once. I remember being there and doing that.' I can picture doing it myself, back in the past, but as you get older, you lose that sense of thinking, 'Right, I've got to be the best here.'

I can remember that occasionally you'd get older comedians – I can't think of anyone specific, so let's say Michael Palin – they'd be in the room with us all trying to be the funniest, but they'd always be very quiet, just looking at us, not part of that melee, and I think Paul and I have now reached that sort of point. Not that we think we've *proved* ourselves to everybody – it's just the desire or need to do that has kind of *gone.*

In the business I work in, everyone agrees on one thing. If you speak to someone who works in management, or is an agent, or in telly, they'll tell you the golden rule: 'Don't ever bother going out for a night with a comedian who's been around for a bit.' Our management don't invite me or Jim [Moir] or Rowan Atkinson to anything, because we're terrible. We're so *dull.*

But I've known Paul for so long that I'm not trying to impress him and he's not trying to impress me. I'm just happy that he thinks I'm good company and vice versa. I suppose that's the beauty of a long relationship. It's like what marriages go through: phases.

The other thing that was important in the show is that Paul and I are of a certain age. No one would have watched it if we were 40. If we were younger, it would be like we haven't earned the right to make a TV show where we do nothing. We would have had to have been funnier. We would have had to address *The Trip*, or do some characters, or have a story like *Detectorists.* And we didn't have to have any of that. And that's been a joy.

When we were doing the first couple of episodes, Paul and I, we are comedians, so we were making sure we tried to be funny whenever the cameras were on. For every 20 conversations we have, maybe one, or half of one, is worth using. But as it happened, we weren't funny all the time – and those are the bits in the show that people found touching and really seemed to enjoy.

But it's not as if we sat down beforehand in some bland BBC meeting room and had that conversation, all of us banging on the desk and saying, 'Yeah, we should go out there and make this touching!' That's a lesson we've been taught through doing the show.

Gone Fishing seems to have become more like *Detectorists* in spirit. I mean, it's always been a bit like that, but it's become that a little more. We're more confident about not necessarily being hilarious and just talking. We're simply spouting whatever comes to mind and reflecting. It's quiet.

If anything, when I watch the show back, we need to do less. Paul and I were looking at the edit, and saying, 'Hang about, we're talking too much.' Stop the talking! There's a tree. There's a river. That's what we like.

Before doing *Gone Fishing*, I've always edited all mine and Jim's telly stuff obsessively. Jim's not bothered, but I edit our work *obsessively*. However, I don't get involved in *Gone Fishing*. I don't know this genre, so what right would I have to start telling people it should be this way, or it should be that way? I don't have a clue.

If we need to lose a minute, I'll come in and they'll ask me about it, but I don't know the beats and the timings of what the factual entertainment lads are doing. I wouldn't have the first idea how to compile whatever you call this show. There's real skill there. With a comedy instinct, I wouldn't know how long people would watch a river going by – should it be ten seconds? Should it be four? I don't know.

If I'm honest, I'm not that bothered about working any more. Four years ago, I'd be, 'We've got to get this, got to push this, got to keep on top of things, maybe we can eke that out.' But after my heart thing happened, the reality was it was a big turning point in my life. I felt that I'd faced up to a very big risk (although it wasn't, as it turned out) and it changed the way I approached my job.

I don't know when or why it goes, because I'm not saying I'm perfectly laid-back and don't care. Conversely, the heart issue also gave me a little sense of purpose. I didn't realise until I looked back, but it's interesting. I've started doing some of the things I always wanted to do.

The recent series of *Big Night Out* have been exactly what Jim and I wanted to do. And it's *nuts*. It's almost back to the Tanita Tikaram stuff we were doing in the 1980s. Nice to think that Paul was there when Jim and I were truly liberated and mad, and he's here now when we're the same.

I've also wanted to do something with football. I'm into football in an almost stupid degree. Football's my number one. But I've never even mentioned football with Jim. If we're drawing things on Man with the Stick's hat, I'd say, 'Oh, let's have Peter Crouch putting a bell in a whatever,' and Jim would be, 'Peter Crouch? Who's that?' So there's never been football in what we've done at all. But I thought to myself, 'I want to do something with football,' and so I did the Athletico Mince podcast with Andy Dawson. I'm just writing an Athletico Mince as we speak – I've reached a point where I'm telling a story about Peter Beardsley accidentally killing Mark Lawrenson's favourite pet rabbit and I do enjoy that.

For *Gone Fishing* Paul was quite clear about when we did a second series, he said we needed to make the same series again, and that's what we did – we've plumped for that. I suppose we could have received a bit of criticism for it being the same in tone as the first. But if I think of myself as a viewer of a series I've loved, there have been some occasions when I've thought, 'What have they changed it for?' Tim and Pru did that – after a few series, suddenly they went to India. And I was thinking, 'Oh, Tim and Pru, no, no, no, no!' What I like is watching the quiet British countryside. I like Tim saying, 'Today, love, we're going to Bodiam Castle,' and then watching the two of them go to Bodiam Castle. So we've plumped for making the same series again.

When *Angling Times* made *Gone Fishing* their telly programme of the year, Paul was *so* chuffed. He does worry about what the serious fishermen will think of him. He says to me – you'll never see it in the show, we cut it out – he occasionally says, 'Oh, that was a terrible cast, we'd better not use that. We shouldn't be using this, we're going to look stupid.' What he means by that is he'd love to emerge from this with the fishing community, who are important to him, thinking, 'He's a good fisherman, that Whitehouse.' So he was really chuffed with the award. I texted him when I found out – I said that's really, really nice; it makes me very proud we are enjoyed by the lads. He just replied, 'Excellent, innit?' I could almost feel him beaming with pride on the other end of the phone. To be honest, he'd rather have that award than any other you can get.

I've just watched two shows from the second series and they're just amazing. The theme of every show in the first series was our hearts, really – that was the theme. There isn't quite so much of that second time around. I said we can't keep doing that. It looks like you're trying to milk it or get sympathy, and no one wants to see that.

When I say I'm not that interested in doing work since my heart scare, I'm being a bit disingenuous, because doing this has been a great job. A *great* one. It's very specific because normally you'd have four months of writing and meetings and angst, whereas – we just go off fishing. So there are still buzzes to be had. Oh man, what a blessing!

I do hope we'll be able to make more of the TV show, but I don't know if we can continue traipsing around British rivers. I suppose we could, but if the BBC are bold enough to say, 'No, I think Paul and Bob just being Paul and Bob are good to watch,' then maybe they'll let us go to America or somewhere like that. Imagine those vistas! So far, we've only filmed the shows in the autumn and winter, so it would be nice to head out somewhere where there's a bit of sunshine. If we go somewhere sunny, I will apply fierce pressure on Paul to make him go skinny-dipping.

But whatever happens, I'm always ready for the call to make another series. I'm ready *now*. I cannot emphasise to you how much. The phone could ring as I'm typing and Paul could say, 'Shit, Bob, they've just commissioned another one, but we have to start today,' and I'd be like, 'Paul, I'm ready *now*. Where are we off to?'

Even my kids quite like the show. They never talk about any of my shows *ever*, never have done – maybe they've looked at them on YouTube or something, but they've never said owt. But they both watched *Gone Fishing* and they said that it made them sad but they quite liked it.

Paul and I are quite proud of what we did. We don't really talk about it much with each other, and it's different to what we've done before, but the people who liked it have *really* liked it and that's great.

Whenever I watch them back, I think they might be amongst the best shows I've ever done.

B: Even after doing the first series, I still don't get many people coming up to me with fishing chat.

P: I get a little bit. What we mainly get is a lot of women. I should probably finish that sentence. We get a lot of women *who watch the show* – probably because their husbands are watching it and they can't get away – and who realise there's a lot in there for them, because it's not really a fishing show. It's about the joy of the British countryside, and the relationship between Bob and me.

B: We're able, by the gift of TV, to present a really lovely relationship.

P: Yeah. Because we can edit it.

B: I suppose every couple at home is thinking, 'Oh, we were once like that,' or 'Oh, we could be like that if things were different.'

P: What they don't realise is that we get on because *we don't live together.*

B: That's the secret.

PAUL

Honestly, it was a half-arsed idea.

We were sitting on the riverbank on a fishing trip together, and I was saying something stupid to Bob – we were both making each other laugh. I think I was pretending to be an Indian naturalist: 'Look, Bob, is a reed varbler!' and for Bob, that is *gold.* He was nearly hysterical. I was like, 'Look, it ain't *that* funny, Bob,' but inaccurate Indian accents are like catnip for him.

Anyway, we were chuckling away and it honestly just came to me in that one little moment: 'This might make a really good TV series.' It just seemed to me like a lovely idea and Bob, he got it straightaway.

I can't remember at what point we came up with the overall concept – a few people chipped in with us when we started to talk about the show more seriously – but I think it was quite clear early on we weren't going to sit down and write anything. The idea had come out of our downtime on the riverbank, and so that's what we wanted the show to be.

We've never sat down and written something together; we've never tried. But I love working with Bob. I love it. And what's really touching is after two series, we haven't lost sight of the root of what we set out to do.

But on day one, with the cameras trained on us there, I did find myself thinking, 'What on earth have I done?'

It's difficult for me to be objective about fishing because I've always had it in my life. Primarily, I go fishing because it gives me nothing but pleasure. But it's also my chosen method of escaping from the stresses of work and all the pressures that come with that. Without getting too pretentious about it, doing the job Bob and I do, fame does invade your life in both good and bad ways. But the one thing it definitely does, without doubt, is change your life.

So having the option to go fishing has always been important to me. Fishermen don't care if you're on the telly – they just want to get on with their fishing, and as long as you're not letting off fireworks or snagging their line, they're going to be more than happy to let you get on with it. Thus fishing has always been my way of getting away from it all, and that's been very precious to me.

So I found myself on day one of filming thinking, 'Oh no, what have I done? Why am I turning something I love into my work? Why have I made my real life into a show?' Because we've never done that before, Bob and I: we've always been characters. 'What have I gone and done here? I don't want these cameras around! And who's this irritating child standing next to me?'

But we saw it through because it became apparent that what we were doing was so genuine and so full of heart that it seemed to override all of those other considerations and concerns.

There's absolutely no artifice in the show. It's pretty much just a record of what we're like together, whether the cameras are rolling or not. For a start, I *genuinely* have no interest in what Bob's saying to me. I really don't! He'll start chatting away, 'Piss off, Bob, I'm trying to watch my float', or he'll be asking me about the brand of vacuum cleaner I've got when I'm trying to cast. *Shut up!*

Obviously, we don't play up to it, but there is now a sort of hierarchy when we're filming, which has become Hapless Bob and Slightly Disgruntled Paul. I do have to be slightly more grown-up and erudite on the show than I am in real life, and I'm not that good at it. Oh, I can do it: I can talk about fish and rivers, and, as the previous chapters might have revealed, I know quite a lot of stuff – you acquire it by years of just being, quite literally, immersed in it. Do it long enough and it basically rubs off on you, even though you're barely aware at the time that you're learning it.

But having that kind of knowledge in your head is a very different beast to standing in front of an audience at home and going, 'Well, Bob, the brown trout is a most fascinating fish. Its natural habitat, the uplands river, is …'

I don't think I'm a good teacher on-screen. I'm not very patient with Bob. He says himself, 'Paul, I'm very easy to tell off,' and I've noticed I do get very snappy with him. A mate of mine watched the show and said, 'Ooh, I saw a very different side of you on *Gone Fishing*,' which was basically half an hour of me hissing, 'Bob! I've already told you!' over and over again.

There's one bit in the second series where Bob asks me about my first kiss, and I tell him and I get back to the fishing for a bit, and then I think, 'Well, I suppose I'd better ask him about his first kiss, too,' – I always forget. At which point, Bob starts making up some name, which is obviously made up. He says he's doing it to protect her identity, I'm saying he's doing it because he's never actually been kissed by a girl, and at that very moment, at that exact point, *I lose a fish.*

Well, the air turns blue, I start shouting, Mortimer! I just missed a fish because of your banal nonsense about kissing some girl when I was nine years old!' Oh, I really went in hard. And there was a little silence. Everything went silent, and I shuffle a bit and go, 'Er, sorry, Bob. Sorry. That was a bit out of order.' And Bob softly says, 'No, Paul. Quite understandable.' We basically covered all the bases there in 30 seconds, didn't we? 'We're having fun!'; 'Your banal nonsense'; 'I'm sorry'; 'No, you're right'; 'Yeah, I am'.

There was also an incident when we were salmon fishing. When you go coarse fishing, you get your pitch and it's yours for as long as you want it. But with salmon fishing, you get in and then you move. It's quite disciplined, salmon fishing. I remember when I first did it I was thinking, 'Cor, it's a bit regimented, this, isn't it?'

There's a kind of unwritten law – a code, a behaviour – that you get on the river and you move down, then you get out and go on to the next bit, so the next person, such as they are, can get in. Even if there's no one else waiting to get on the river, there's a kind of notion that a pool is best fished down two or three times quickly than once slowly. After all, you're not putting bait in to draw the fish up – they don't feed anyway – so as you can't lure the fish to you, you're better off moving to actively find them.

Anyway, Bob and I are on the river, and we see a salmon move upstream – up beyond Bob, but the sort of one you could nick out and have a go at – and then almost immediately we see one move downstream. And I said to him, 'Right, you go down and catch that one AND BY THE WAY IF YOU'D WALKED DOWN THE RIVER WHEN I TOLD YOU TO WALK DOWN THE RIVER BEFORE, YOU'D BE ON IT RIGHT NOW!' Watching it back, there's anger in my face – real anger. I was standing in the edit suite, horrified, thinking, 'Jesus, calm down, Whitehouse! The bloke's barely been fishing in his life!' What a *horrible* man.

So it was nice to see Bob say I was a good teacher. I think that when we're alone and the cameras haven't been there, I'm much better then. It's nothing as tense as I am on the show. I'm a lot more patient.

Speaking of my young apprentice, it's noticeable to me how, in the course of a few years, Bob has really come on. He loves fishing, he gets totally absorbed in it, and his technique has come on in leaps and bounds. Very early on, I noticed he's got a real affinity with it all. Obviously he fished as a child, but most important of all, he'd always had that burning desire to do it again. He would probably be a bit dismissive of me saying this, but he's actually really quite a thoughtful fisherman. He seems to have a sort of natural aptitude or affinity for the water: quite aware of trying an area a little off the current where a fish would lie, or spotting a promising bit of still water.

Yes, it took a near-death experience to get Bob and me on the river together, but now he's doing it and he couldn't be happier. A lot of people do ask me whether I'd ever go fishing with Jim if he nearly died, too. I suppose I'd have to now, otherwise it would look like favouritism.

But the thing that's really impressed me most about Bob is that all he wants to do is fish. We do all those introductory bits and the drive-bys, but the reality is the two of us, once we're out there, *we just want to fish.* In my book, that makes Bob a proper fisherman. He's actually more obsessive about it than me: I get to a point where I'm ready to call it a day, but Bob will always want to fish on.

If anything, having your time out on the water being limited by having to make a TV show amplifies that obsessional element of wanting to be left alone just to fish. Bob and I are happy to be fishing all day. Even when we lose the light and the crew begin to pack up, we'll still be standing in the river, fishing. Then afterwards we all go to the pub, usually with the whole crew. It's a joy.

The cameramen on the show are all cut from quite intrepid cloth. Often, they'll be standing still, waist-deep in a cold river for six hours, so while the shows always look very beautiful and placid, it can be pretty tough on them physically.

I remember when we went out after the bass off the Isle of Wight – the sea wasn't too bad, but it was a bit rough. Bob was terrified he was going to get seasick so he'd taken loads of anti-nausea pills; I hadn't taken any, so was pushing my luck. But the two of us are doing something – we're fishing or we're being filmed – so we're focused and it distracted us from the swell and rise of the waves. When you're doing stuff, concentrating on something else, it kind of takes the edge off the choppy water. But there was a moment when our producer, Nicky, got really sick – I recall she honked seven times – and there's no question of us going in. Everyone knows it: we're *not* going back in. So sometimes the poor crew have to sit there and suffer for our art.

In fact, that's one of my favourite episodes so far – the one where we go after sea trout and out to sea in search of the bass. That felt like a proper journey and it seems like a great big open episode. But the barbel on the River Wye, that might be my favourite episode overall. Those barbel were *monsters*. It's such a weird fish with its fleshy feelers, it really is specifically designed for its environment. We invited a vicar on to that episode to talk about God. I asked her, 'So did God design the barbel for that environment, or is it quite the opposite? What does that barbel represent: God's work or evolution?' When you look at that fish, there's the big question, writ large! I asked the barbel the same question when it came out of the river, but it didn't give me any answer. I might never know if there's an intelligent higher power, but what I do know is I really like that episode. And we dedicated it to my late father, who I fished that river with when I was a kid.

Quite a lot of people probably don't realise that Bob and I have been fishing together between the series – and I'm so pleased we've done that. It hasn't just become a work thing, because I was very conscious of that when we started.

It was Bob who initiated it. We went pike fishing with John Bailey and Bob caught this monster pike – he was so pleased with himself. Then Bob took us back to the Derbyshire Wye, a river he really has a soft spot for. We stayed at a beautiful hotel,

Paul: Bob got lucky. In that I introduced him to the great John Bailey, who introduced him to this huge pike …

but technically, it was a difficult style of fishing for Bob, who is a relative newcomer to fly fishing. I winkled a few fish out, which impressed him, I think.

See! Without the cameras and without the dialogue, I *do* know what I'm doing. Weirdly, when they're there, my technique goes to pieces. But when I'm just fishing and I'm allowed to just concentrate and fish, I know I can winkle a fish out.

I can take a lot of pleasure in doing the fishing for the show, but there is a lot of pressure to catch a fish. People say, 'Wow, you get to do that – fish all day in some incredible locations,' but it's also a proper day's work. This is not just Bob and me mucking around! You've got to get a bloody fish on one particular day, and when there's a crew of half a dozen, you can't go, 'Well, we'll come back tomorrow, and if we don't have any luck then, we'll come back the day after, too.' I do try and get a bit of extra time built in – I tend to go for pleading for another day's worth of filming – but it's not always practical and a camera crew doesn't come cheap. So does the ticking clock stop the fishing for the show becoming a completely joyful experience? In a word: yes. In three words: yes, it does. A bit.

I don't know how important the actual fishing is to the people watching the programme, but I think it's *really* important. Without people realising it, it's absolutely vital. Yes, there's this daft obsession that has actually taken us there, but we need to get some results – otherwise it's just us aimlessly wandering around. Without the catch, it wouldn't mean anything at all. Trying to get those fish in our nets gives the show a sense of jeopardy –

even though it's not the main jeopardy, because with our heart problems either one of us could drop dead any second.

But it also means you get some thrilling moments when you get a last-minute victory. For example, we caught a good trout on the Derbyshire Wye, right near the end of filming. When we did the show about tench, we caught a big tench on the very last shot just as the crew were packing up, and that was *fantastic*. The anticipation, the countdown, the creeping feeling that we'd failed, the slight glimmer of hope that we might change our fortunes if we stayed on another minute, and then *BANG*! Thankfully, the crew had the drone up doing the final aerial shots, so we got it on film by the skin of our teeth. That was an incredible feeling. A triumph! And I caught a huge sea bass on *literally* the last cast of the session during the episode when we went after both sea trout and bass. A lot of people overuse the word 'literally' but that bass took on literally the last cast before the boat was due to head for shore.

Bob loved the first part of the episode when we fish for sea trout from a kind of pontoon on the bridge pool of the Hampshire Avon at Christchurch. He often asks me how you go about getting on the pontoon, because he's thought of taking his wife, although based on prior experience, he'll have to get someone to set his gear up for him before he goes. I tell him again and again how he can get on the pontoon, but he never remembers – it's usually a week or so before he asks me again, for the 40th time – so, Bob, I've put the following bit in, solely for you to clip out, pop in your wallet and look at next time you want to know.

> You have to ask that Jason, the bloke who runs the whole of the Royalty Fishery rig. It's on the outskirts of Christchurch, and it's got everything: sea trout, salmon, and all the coarse fish. So phone Jason up and take your rod and a bucket of maggots. And then you're off. Now please stop asking me. *Please.*

But we might be slightly building a rod for our backs now, simply by the fact we're trying to make a programme that often builds to a really dramatic denouement. We've almost set our own template for the show, by the simple fortune that we've had with a few shoots where we didn't catch a great fish until right at the very end. And that means we never want to jack it in if we're having a bad day, because experience has taught us if we catch something in the dying moments of the shoot, it's an even bigger thrill for the audience and for us than if we were to catch loads at the start. The thrill of thinking you're not going to do it, but then doing it, is most definitely not what regular fishing is about.

If we do catch a fish, we do. If we don't, we don't. We might have two episodes out of six that end in disappointment, and that's okay, because two days out of six regular fishing days might end in disappointment. Despite your keenest hopes, fishing doesn't always have a happy ending.

One thing Bob was slightly concerned about when we started the show is that fishing is not very politically correct. But the only people who've moaned about anything is the anglers, all saying, 'Why did you keep that bass you caught? That's a disgrace!' They love the fish so much! But in a way, they've got a point – sure, the bass is a big sea game fish and most fishermen would keep them, but maybe we should have put it back to set a good example. So hats off to those anglers for being on our case! Mind you, it didn't surprise me that anglers were first in the queue to have a pop. The reason we kept that bass is because most people have eaten bass at some time. Oh well, you can't please everyone.

I was adamant that the later series should be similar to the first, but as we keep going, who knows what it will become? Bob's always saying, 'Oooh, can we go to Montana, Paul?' Big Sky. I mean, for my part, I've always fancied Slovenia – it's very beautiful and there's supposed to be extraordinary trout fishing and grayling. Iceland is always a good one, too. And we haven't caught a grayling, haven't caught a salmon and haven't caught a big pike, so …

If it remains about the two idiots on the bank and we can capture that raw extraordinary beauty where we fish, then sure, *Gone Fishing* can be filmed anywhere. So long as the heart of the programme remains what it is – which is Daft Bob in a fishing situation with slightly less daft me, and we enjoy the quiet, the solitude, the peace and the little bit of humour that comes out of that.

After all, people – for want of a better expression – hold Bob and me in a lot of affection. 'Oh, them two together! Oh, that's nice! This will be fun!' Then they watch a bit of it. 'Oh, no, wait – they're not very well.' I don't even mean that last bit cynically. People see that, 'Oh, they're getting on with it and neither of them are being too po-faced about it,' and we can mock ourselves while at the same time taking the heart stuff quite seriously. The show sends out a nice, positive message to people with problems, and it says if we can do it, then you can, too. Look how beautiful Britain is. Go and see the country that lies on your doorstep. Look how beautiful life is.

The show has a few different things that people enjoy watching: it's the heart jeopardy, it's two old mates and you're asking, 'Will they be getting on all right together?' It's my slight grumpiness with Bob. There's the teaching, there's the fishing, there's the splendour of the English countryside, it's filmed wonderfully, and there's the gambling element of whether we'll manage to get anything from the river. And all those things, they all seem to come together. And despite all those things, despite being on the river for five hours, despite all those elements, we're not trying very hard and that's the key thing.

We're just two idiots being friends. We lost each other a little bit but a trauma has brought us close and we're saying, 'Hey, *anyone* can do this.'

It's not that hard.

You just have to go fishing. And preferably with someone as funny as Bob … That's the really difficult bit, because they're aren't many like him.

P: And when we go fishing without the cameras, we can be really funny, can't we?

B: Yeah! It's quite nice, because of the privacy of it. We can say whatever we want!

P: There is a bit of that, isn't there?

B: Say whatever we want, Paul. And we say some awful things. We do, Paul. *We do.* Some *terrible* things.

P: Ha ha ha ha! We do!

B: Yeah. Best that they're left out.

CHAPTER 9

THE FISH – A GUIDE TO OUR RIVER PALS

Paul: Hands off, Bob!

PAUL

Water is very mysterious. Both running water and still water fill me with a sense of slight wonder. There's another hidden world down there, filled with strange creatures that are both a key part and entirely separate from our own world on the surface. Fishing is the means by which we can break through the barrier and enter that other place.

The angler's float is the bridge between these two worlds – it's effectively a doorbell through which humans and fish connect with one another. And when that float disappears, I love that moment above all others. For me, it's like the reaction on a baby's face when you hide something and then bring it back again – the moment the float disappears, something magic has happened. It's making contact with an alien species.

And what a species fish are. The colours, the charm, the personality, the spirit – too many people dismiss fish as being little more than cold-blooded emotionless muscles that swim, eat and get startled. And there's some truth in all those things. But fish are so much more than that.

They have individual characteristics, unique abilities and fascinating histories. They have shown they can learn and adapt to changes in their environment. And they've been around far longer than humans. While our ancestors were scrabbling around in the dirt and banging rocks together (see Chapter 5 and the story of the legendary fishing monkey-man), fish had already finished their evolutionary journey. They reached their full potential long before we did (and that's only if we have – it would be a depressing reveal if it turns out the humans of 2019 are as good as we're ever going to get).

And consider this. With an estimated 3.5 trillion fish on earth, they outnumber humans. They outnumber birds. They even outnumber trees. I don't want to get all New Age on you, but perhaps it's worth taking a moment to consider which species this planet *really* belongs to.*

* Yes, the answer would be bacteria, but that's not the point I wanted to make here, so I've just ignored them. Same with insects. Look, this is my book and I'll use any data I find how I want.

BARBEL

Barbus barbus

aka Sergeant Barbel – the Bottom Hugger

> 'The barbel affords an angler choice sport, being a lusty and cunning fish.'
>
> *Izaak Walton,* The Compleat Angler

The name barbel comes from the Latin *barba*, meaning 'beard', and it's so-called because of the four fleshy bulbous whiskery bits that hang off its lower lip. As it's called *Barbus barbus*, it seems to be called Beards beards, which is the name I would prefer it to be called at all times.

These barbules are sensitive to the slightest movement, similar to, I imagine, the human testicles.* They also contain taste cells, which are not a feature of the human testicles, for which I am grateful. The barbules mean the fish can sniff and taste the food it encounters before putting it in its mouth.

---

* Bob: In his dotage, Paul has frequently been found sitting around, imagining testicles.

The dark-brown, green and golden-bronze barbel was also known as 'the pigfish' in England many years ago, due to the way they snuffle around for food on the riverbed in the same way a pig snuffles and truffles through the mud, looking for disgusting wet snacks. Nowadays, it's often referred to as the Prince of the River. I presume they mean 'Prince' in the royal sense, and not the late funk/pop music performer, but I'm not a fishing expert, so I'm willing to be corrected if anyone knows for sure.

Sitting near the top of the food tree, the barbel has very few predators – many scuba divers report that barbel don't swim away when they approach and instead just hang out because they've never had to fear anything in the water before – but the reintroduction of otters has occasionally proved problematic for the fish. Case in point: the murder of the Big Lady.

The Big Lady was the UK's largest barbel. She lived in the River Ivel in Bedfordshire and weighed over 20lbs. The Big Lady was big, and a lady, and beloved by the local fishermen. But on a dark night in July 2015, an otter dragged the Big Lady out of the water, tore her throat out and left her to die. No one deserves to die like that, especially not the Big Lady.

'I heard this sickening crunching sound as I approached a reed bed with my torch,' said the angler who stumbled onto the crime scene, who wishes to remain anonymous. 'The otter vanished and I looked down to my horror as the huge fish – which I recognised instantly as I've caught her four times before – lay there dying. There was nothing I could do.'

At the time of writing, the murder remains unsolved.

The UK's previous biggest barbel, the Traveller, was also supposedly murdered by an otter a few years before, but more recent evidence suggests he may have died of old age and a passing otter then simply desecrated his delicious corpse. For many years, catching a fish over 20lbs was the Holy Grail of barbel fishing, and in 2011, the Traveller was the first barbel to meet the criteria.

In whatever circumstances he died, the Traveller is now continuing his travels – this time to heaven.

From the 1960s to the 1980s, the US Navy had a nuclear submarine called the USS *Barbel*, which was decommissioned and then used in the 1995 film *Crimson Tide*. It stars Denzel Washington and Gene Hackman, so why not rent it (or whatever modern people do these days) and shout, 'That's the USS *Barbel*!' at your unprepared family the moment you think you see it?

SIZE

There's quite a broad range of sizes in the barbel universe. From 10 to 120cm in length, and ranging from 1 to 12kg in weight, there's a barbel to fit every pocket and every budget.

CHARACTERISTICS

The barbules, mate. It's got barbules. I've *just* been telling you about them. Come on, man, you've *got* to pay attention. It's just a waste of everyone's time otherwise. *Come on.*

Barbels like to wander around a lot, and some have been noted to move several kilometres in a single day.

WHERE IT'S FOUND

As Izaak Walton once said, barbel prefer 'the strongest swifts of the water, and in summer they love the shallowest and sharpest streams'. They search out clear water with a hard river bottom covered in silt, where their awful foodstuffs can be rootled up. In winter, they semi-hibernate in big shoals in deep water, only coming out on mild days to grab a bit of snap.

DIET

Barbel have quite catholic tastes and will devour insect larvae or nymphs, worms, silkweed, crustaceans, even small fish. Some anglers swear they're easy to take if you bait your hook with sausages. As far as I am aware, this is the only fish that is known to eat sausages. They are also very partial to luncheon meat. That's not very heart-healthy, guys!

LIFESPAN

Barbel live up to 15 to 20 years. It's not a long life, but the possession of those fleshy barbules mean it must be one constantly filled with wonder.

SCARCITY

Now abundant, the barbel is very firmly in the least concerned category. That said, pollution over the 19th century affected their numbers greatly, and for much of the 1900s they looked to be on the verge of extinction in Britain's waterways. Since the 1980s, their numbers have soared, thanks to restocking and reintroduction.

HOW TO CATCH A BARBEL

In the right conditions, barbel can be relatively easy to catch. But there is well, a catch – they do like to fight tooth and nail (although they do not have nails, they do have hard and strong little teeth, like the rest of the carp family, but they are pharyngeal which means 'in the throat').

Experienced fishermen claim that weight for weight, the barbel fights as hard and as long for its freedom as the salmon. So you'll need a strong rod and a strong line – 10lb and up. Nowadays, most barbel fishers would use a hair rig, but if, like me, you recognise those words but wouldn't know how to set one up if your life depended on it, you might wish to stick to more simple methods, including prayer.

The Hampshire Avon, the Trent, the Severn, the Loddon and the Herefordshire Wye are all celebrated barbel waters.

PIKE

(Esox lucius)

aka Ron Pike – the Boss

> 'The tyrant of fresh-water-fish…it is the general opinion that no other fish will associate themselves with this water-tyrant, for he always swims alone, and is the most bold and daring of all our fresh-water fish, knowing no other pleasure, as we conjecture, than prey or rest.'
>
> *Charles Bowlker,* The Art of Angling *(c.1746)**

You will never know true terror until, one quiet English morning by the riverbank, the morning mist hanging softly over the meadows as the hazy sun begins its silent climb across the dewy woods and flowers, a pike takes your hook or lure without warning and almost drags you screaming to your death.

The Northern pike carries with it an air of dead-eyed menace, like the ringleader of a 1940s razor gang, terrorising café owners

* Get it right, Charles; everybody knows the big pike are all female!

who have fallen into debt through gambling. It lurks in the dark, shadowy waters like it's trying to nick your mobile phone; skulking in the shadows of riverbanks, waiting to rush whoever's passing and bite them on the face. But is this aggressive reputation deserved?

Yes. Pike are horrible. But they're *spectacularly* horrible! Fossil remains show the pike has changed very little over the last 20 million years: it's a truly prehistoric fish.

In British waters today, pike are the apex predator. Feared by everything else in the river (including other pike), they have a hardman reputation to uphold. So if you come near them, pretending you're the big man and giving it all that, they're going to remind you who the river's guv'nor is with a swift and sudden pop (by which I mean 'eat you if you're smaller than they are').

Their name comes from the medieval weaponry, the pike (which in Middle English means 'point' and may have been derived from 'pique', which is the French for 'spear') – it was thought they looked similar, being long, slender and with a horrible sharp bit at the business end.

Pike have rows of top teeth which are as sharp as razors and point backwards, while on the bottom are teeth shaped like needles. It's basically got a knife drawer for a mouth. And once those jaws are shut, anything inside is pinned without hope of escape.

Olive green, with a paler belly and covered with light and dark spots (which are as unique to pike as fingerprints are in humans) which help them merge into the dappled depths of water, the pike is an ambush predator. It will lie completely motionless for long periods of time in a concealed, gloomy part of the river, waiting for its prey to swim past. Waiting. Watching. Psyching itself up. Waiting some more.

Once a hapless victim ambles by, the pike bends and then uses the tremendous acceleration given by its large dorsal, pectoral and anal fins (which are all positioned well back on their bodies) to shoot out and strike. All the fish will see is a shadow at the last moment. And that shadow means death. They also have an anti-coagulant substance in their mouths that acts like a blood thinner and inject their prey with warfarin to ensure a lot of bleeding takes place. It's a killing machine!

Pike are the subject of one of the most famous poems written by Ted Hughes. A very keen angler for pike, Hughes recalled walking down on a hot day to where he used to fish: a large pond which was very deep in one place. In the sun, he saw 'something like a railway sleeper lying near the surface' – a huge pike. Remembering his pike fishing days years later, he wrote one of his most celebrated poems. It was called 'Pike', which suggests Ted might have been a master of iambic pentameter, but he was not particularly bothered about titles.

Similar literary heavyweight DH Lawrence also referenced the pike in a poem called 'Fish'. Has a classic novelist ever put pen to paper about the way you conduct yourself in public? Has a Poet Laureate ever written a poem about *your* sag belly, malevolent grin and submarine delicacy and horror? No. So don't think you're better than a pike until they have.

SIZE

Mature pike are on average between 16 and 22 inches long. Interestingly, male pike rarely get above 7kg, and anything bigger than this is usually a female. But with enough fish to eat, both sexes will keep on growing. In the 19th century, pike were huge by today's standards, since they were able to fatten themselves on the constant annual stream of salmon coming upriver. With the decline in salmon, truly giant pike have also disappeared.

In 2013, a pike skull was found at the River Cherwell that contained 700 teeth and was a foot long. This meant, when alive, the pike could have been around 54 inches in length, weighing around 60lbs and bigger than any pike ever caught in Britain (although a lot of anglers doubt this, saying it's impossible to accurately work out a fish's height and weight from just seeing its skull). It's amazing to think it was out there for 20 years and no one ever recorded seeing it, even though it would have looked more like Nessie than a fish.

The largest pike ever recorded worldwide was a mammoth 92-pounder. It beached itself in shallow waters in Ireland in 1832, where two men ran over and beat it to death with an oar. Ah, the eternal romance of fishing! No wonder catching a pike of 10kg or over is considered the qualification of a master fisherman in Finland.

CHARACTERISTICS

Like East End villains, pike each control their own manor. They're incredibly territorial, and the bigger the pike, the bigger area it will control, chasing off any other muggy pike who come along and have the nerve to try and wet their beaks in the nice little earner they've got going on. What a liberty!

That said, there is a school of thought which says they're not actually territorial so much as actively scared of being eaten by one another – so small ones tend not to hang around areas where the bigger ones have settled.

And if we're talking about characteristics, then it's worth mentioning the pike has the most awful face of all freshwater fish. And, as a species, fish aren't exactly known for being pretty.

WHERE IT'S FOUND

Pike are found in lochs, lakes and sluggish streams: anywhere there's a ready supply of fish for them to hurl themselves at and devour like something from a horror film. They also inhabit some faster-flowing rivers but will stay well out of the main current. They like clear water (so they can see what's passing by) and weeds (so they can hide) so reedy shallows and dark spots under overhanging trees and under the bank are always worth a try. Rule of thumb: if it looks like somewhere you wouldn't want to stick your hand into, it's probably the perfect home for a pike.

DIET

Pike will eat anything that moves. Smaller fish, frogs, insects, rodents … and while it helps if their prey is smaller than them, this isn't something they're unduly strict about. They'll happily pick off entire lines of fluffy little ducklings, they've been known to drag adult ducks down by their legs, and occasionally have a go at little dogs that get too close.

They're also famous for being voracious cannibals. It's especially common when other food sources are scarce. Dead pike have been found suffocated with the body of identically sized pike they've tried to eat still stuck in their mouths: Ted Hughes's poem 'Pike' (still a lazy title) mentions seeing this very sight.

LIFESPAN

Ten to 15 years is the norm, but pike living for 25 years have been recorded.

SCARCITY

Pike are common in Britain but, as an apex predator, not as abundant as the fish species that they feed on. For years, it was said they were in every English county except for Cornwall, but the growth of the internet has revealed lots of fishermen catching big pikes there, so ignore this inaccurate slur that suggests pike hate Cornwall. They don't.

There are also some monsters lurking in those big still clear lochs in Scotland – Loch Lomond is renowned as one of the best pike fishing waters in Britain.

HOW TO CATCH A PIKE

First off, pike fishing can be a challenge for the beginner. Really big pike are thin on the ground: the fish can be hard to locate and they're extremely sensitive to outside barometric pressure; the really efficient hooks used to catch them are more complex and even when caught, they put up a real fight.

You'll need a carp hook, treble hook or circle hook – the pike has a very bony mouth, ridged with hard white bone, so an ordinary hook won't usually be able to get much purchase. Any one of those hooks loaded with some dead bait (mackerel is a favourite) or a nice little spinner will get the pike's attention. Anglers used to swear on the efficiency of live bait for pike, but these days, the use of live bait is rather frowned upon in polite fishing society.

Pike are large and powerful, so you need a strong line. You'll also require a wire trace between your lure and the line because otherwise its sharp teeth can cut it. And if it gets away, the pike promises you this: it will come for you and it will finish the job you started.

Two things to be aware of. First, because of the way they take their prey, it's very common for them to take the hook deeply. If fishing for pike, you need to take along a pair of long-nosed pliers or a disgorger in case you need to work the hook back out. Remember to be careful of its deadly teeth when you're removing anything, or it will be returning to the river with half of your hand as a packed lunch.

Secondly, despite their loutishness, they are an unusually delicate fish. Care must be taken both when handling them on the bank and when putting them back into the water. Like all East End thugs and geezers, they become very sensitive the moment you hold them in your strong, muscley arms.

ROACH

(Rutilus rutilus)

aka Chester Roach – the Gaudy Merchant

'There is a peculiar charm in roach fishing, as is evidenced by the number of devotees. It is difficult actually to define this charm, this fascination, but once infected one remains a roach fisherman to the end… The roach indeed must have been born under an unlucky star. Its pursuers are keen and number many thousands.'

Edward Ensom, Fine Angling for Coarse Fish *(1930)*

To many fishermen, the roach is the most charming of the coarse fish. And what a fish! The upper part is grey-green tinged with blue; the abdomen is white; the sides a silvery white; the fins bright red, with yellow and brown tinges. You might think that would be enough for a small creature. NO. It tops off this look with golden eyes and a red pupil. A tiny, shining little rainbow of a fish, it could easily be something from an aquarium. I mean, it's not subtle, but it's not over the top. They're gorgeous, they really are.

A member of the carp family, for years the roach was thought of as such an easy fish to catch, it was known as 'water sheep'. Fascinatingly, it seems that decades of intensive angling taught the roach to be more wary, and the modern strain of roach you get today are rather shy, cautious and crafty.

The roach is very similar to the rudd and dace, and it's easy to confuse them, and quite hard to tell them apart. As a rule of thumb, rudd are more yellow-greeny-golden than the roach and have an upturned mouth, while the dace lacks the roach's colourful eyes. Even the fish themselves find it hard to work out what's what – roach regularly cross-breed with bream and rudd, the dirty rascals.

It's hard to put a finger on why anglers love the roach so much – it's just a fact. As Mr Crabtree puts it:

> No other fish has so many followers…Once the spell of roach fishing has settled on a man, no other fish can draw him from his love. He is a roach fisherman for life.

SIZE

Roach are small fish, with most in the region of about 30cm and weighing under a pound. Anything over 2lbs is unusually large; specimens up to a maximum of 4lbs have been recorded but are rare, and have all been found in still-water lakes in the south of England, where food supplies have been particularly rich.

The current British record for a roach is 4lb 4oz, caught in an undisclosed location in Northern Ireland.*

CHARACTERISTICS

Roach have a slightly prominent upper jaw and a small mouth, which gives them a natural duck face – like the over-exaggerated, open-mouthed pout that people often do when they take photos of themselves in bathroom mirrors on a night out. We propose that this should be renamed 'roach face' in honour of the plucky roach.

WHERE IT'S FOUND

The common roach is found over most of Europe. It inhabits lakes, ponds, fast-flowing rivers, slow-moving rivers and canals across England, Scotland, Wales and Ireland.

DIET

The roach's usual diet is invertebrates – like insects, grubs, worms and larvae – and plant material, like weeds. It will feed at any

* Bob: Never give away your prize fish waters, Paul. They'll be immediately inundated by glory-hunters, won't they? I mean, the first rule of Caught A Big Fish Club is 'Nobody Talk About Caught A Big Fish Club'. That's a lesson I learned on day one, Paul. *Day one.*

Paul: Was that the same 'day one' that you learned to bang on endlessly about your big pike?

depth and will eat pretty much anything – their adaptability is why they're so successful in all waters.

LIFESPAN

In the warmer climes of southern England, the roach can live up to 18 years. Around the rest of the country, it has a lower life expectancy of just 14 years.

SCARCITY

Roach numbers are abundant. In most of the places it can be found (which is almost everywhere), the roach is the most numerous of fish. Until recently, it was the most popular fish targeted by anglers in Britain, but they've lost that slot to the carp. A cause for celebration if you're a roach. Interestingly, it tolerates organic pollution better than many fish species, and the roach is often the last fish to disappear from polluted waters.

HOW TO CATCH A ROACH

Use a fine line with a small hook (size 20 or 22). The best baits are bread and maggot or caster, and as roach gather in shoals, it's worth ground-baiting the area little and often before casting.

The bite of the roach can be like lightning, and if you don't set your hook quickly, he'll be straight off with your bait. In still waters, the bite can be almost imperceptible.

TENCH

(Tinca tinca)

aka Weary McTench – the Bollard of the Lake

'The tench is for many the symbol of the new season. It is, perhaps, more than any other the totem fish of summer. How many early mornings, translucent twilights we can remember when the tench have set the reeds swaying, and finally our floats have wavered, dipped and dithered, and then slid away.'

Bernard Venables, Mr Crabtree Goes Fishing *(1949)*

The tench is – and I'm going to use the phrase 'the iconic fish of summer' again – the iconic fish of summer.

It's so fixed in most anglers' minds and lives because it's the fish you always fish for on the first day of the coarse fishing season. The 16 June is always a magical day for fishers. In our mind, in our perfect scenario, you'd have a little red quill next to a lily pad in a great big millpond lake, and the bubbles come out, and the float lifts in the weird way it always does when a tench takes it.

It sort of slides away, and then out comes this rather beautiful, rather strange fish. A day like that, it's probably happened three times in history. But even if you can only get close to a day like that, blimey, that'd be a good day!

The tench is a really hard, tough old fish, which looks rather like a carp that's done a lot of cardio to get back into good shape. It's a beautiful, deep olive, gold, green colour with a very vivid orange eye.

I know blokes who have fished all over the world for all sorts of incredible fish and if I ring them up and dangle a tench-fishing session in front of them, they'll go for it in a heartbeat. But it's not always been that way. The Roman poet and teacher Ausonius referred to tench as 'the peasant's food' and 'the solace of the common people'.

Amazingly, the tench has had a sort of emergence within youth culture in recent years. To the young, a tench is a man who knowingly goes out of his way to have sex with his friend's ex-girlfriends. We've come a long way since Ausonius.

SIZE

Tench are not huge fish – very few exceed 60cm and most come in considerably smaller than that. A specimen would be anything over 7lbs, an unusually large tench around 10lbs (only a few weighing in double figures are caught over an entire season, so they're not at all common), and the English record stands at a whopping 15lbs.

CHARACTERISTICS

While part of the carp family, the tench is unusual in the fish world as there is only one single species of tench (*Tinca tinca*) and no subspecies. The scales on a tench are incredibly small and embedded deep in a thick skin. This retains the fish's natural slime, which makes it feel very slippery when handled by human hands. For some reason, the presence of this thick slime led people over the centuries to believe the tench had some sort of mystical healing properties for other fish (I have no idea why coagulating slime should suddenly be accepted as a medical degree, but the past was a different place). It was claimed that if a sick fish rubbed against the tench's 'natural balsam' (in the words of Izaak Walton), it'd be cured. This folklore belief gave the tench its long-standing nickname of 'the Doctor Fish'.

The 18th-century letter writer Moses Browne wrote about how pike specifically wouldn't attack tench, but instead went to them for help:

> The Tench he spares a medicinal kind:
> For when by wounds distrest or sore disease,
> He courts the salutary fish for ease;
> Close to his scales the kind physician glides,
> And sweats a healing balsam from his sides.

The reality is the tench's slime most likely keeps it free from infection in the same way that all fish slime does for its wearer, but where's the fun in that? They share a characteristic with urban pigeons in that you rarely see a very small one.

WHERE IT'S FOUND

Tench are found all over Britain, but are less common in Wales and Scotland and are never found in fast-flowing rivers. Still, deep waters (like lakes and canals) with a lot of vegetation are very much a tench's idea of paradise. 'He thrives very indifferently in clear waters,' claimed the *Sporting Magazine*, 'yet delights to feed in foul ones.'

A shy fish, they move about in small shoals, but don't make themselves very obvious: often the only sign they're there at all will be small bubbles coming up to the surface when they feed.

During the frosty times of year, tench are even harder to find as they dig themselves down into the mud and go into a sort of hibernation. Even if the lake they're in freezes over, it's believed the state they're in (although exactly what's occurring is a mystery) slows down their metabolism enough that they neither require much food nor oxygen to stay alive.

DIET

Snails, molluscs and aquatic insects and their larvae make up the tench's diet, along with a small amount of vegetation. You've got to have your greens. They're much more laid-back feeders than their cousins, the carp.

LIFESPAN

Fifteen years seems to be an average, but some sources claim tench can live up to 30 years.

SCARCITY

Common, but the golden tench is much rarer. Cultivated as a colourful strain of the first tench that came into Europe for ornamental ponds, one is occasionally spotted roaming free in the British waterways. They don't last long due to their bright colour acting like a beacon to predators (the Golden Arches of McDonald's fulfil much the same function).

HOW TO CATCH A TENCH

Dusk and dawn is when the tench like to get together for a feeding frenzy. They're primarily night feeders who love the maggots, but will also go for a bit of sweetcorn or a boilie if they're in the mood. Bread and worms are also very effective.

You need a strong line for a tench, because when they're in the mood, they don't want to come quietly. I call it tench warfare. You may not like puns, so you might not wish to join me in so doing. They'll fight for all they're worth, using their thick, paddle-shaped tail to try and get away.

TROUT

aka Raymond Trout – the River Detective

> 'The angler's finest quarry … if trout fishing is any means to be had, take it. It is the angler's ultimate pleasure.'
>
> *Bernard Venables,* Mr Crabtree Goes Fishing

There are three major varieties of trout in the UK: the brown trout (*Salmo trutta*), the sea trout (*Salmo trutta morpha trutta*) and the rainbow trout (*Oncorhynchus mykiss*).

The brown trout is native to Britain, the sea trout is found all over England, Scotland and Wales (where it's also known as the sewin), while the rainbow trout was introduced from North America in 1884 – but look, they're all *great.*

They've been celebrated for centuries – the Ainu people, supposedly the original inhabitants of Russia and Japan, believed that the earth rested on the back of a giant trout, and whenever it flapped its tail, that's how we ended up with earthquakes.

Franz Schubert wrote a song called 'The Trout' in 1817 and his 'Piano Quintet in A Major' is also better known as the 'Trout

Quintet', but following straight on from that story about the massive trout with *the entire world on his back*, it's hard to be that impressed. I should have swapped the order.

Trout are a single species, but no other fish has as much variation within their species – there are so many different genetics at play in the trout family that they're more diverse even than the human race. This is due to isolated populations of trout developing over centuries without ever coming into contact with the others. Some brown trout in Scotland have inhabited the same isolated lochs continuously since the last Ice Age, cut off from the rest of their species.

Rainbow trout and brown trout share a common ancestor, but some 15 million years ago, they divided into two isolated groups, so while they're related, they're no longer as close as they once were.

Similarly, a sea trout is just a brown trout that at some point forked off (most likely, there was a lack of food in the location it found itself) and adapted to the salt water of the sea as well as its usual freshwater habitat, allowing it to migrate even further out of its native river and find waters with much richer amounts of food. Though there is a school of thought that says it might have been the other way around. Make up your mind, scientists; why can't you stick to your existing half-baked theories like religion does and stop mucking about with everything?

So basically, these trout have evolved in their own very narrow and specific gene pools, allowing them to adapt perfectly to their habitat. Just like what was rumoured to happen in remote rural villages in the Forest of Dean in the 1920s, where entire generations of strange-looking farm boys were born with uncanny strength, massive hairy hands that could lift two piglets at once, and a life expectancy that ran out as soon as they hit their twenties.

Unlike the people of the Forest of Dean in the 1920s, the trout avoid inbreeding by using smell and visual cues to select a mate – the healthier the cues, the less chance the fish has problems stemming from too narrow a genetic stock.

The trout was first mentioned in AD 200 by the Roman author Aelian, who wrote of 'fish with speckled skins' and the brown trout has been celebrated since then. It is the subject of countless fishing books and the granddaddy of them all, Izaak Walton, was also a fan:

> The Trout is a fish highly valued, both in this and foreign nations. He may be justly said, as the old poet said of wine, and we English say of venison, to be a generous fish: a fish that is so like the buck, that he also has his seasons; for it is observed, that he comes in and goes out of season with the stag and buck.

The rainbow trout was first introduced to British waters in 1884, when Atlantic crossings and new techniques in the storing of fish eggs meant it was possible to successfully introduce them to a new habitat thousands of miles away. Yes, they're an invasive species, but they're a thing of absolute perfection: a rainbow, a literal rainbow, of silver, purple, pink and blue.

Rainbow trout is the primary form of farmed trout in the UK, and you'll see it on countless restaurant menus. While these farming methods are low-impact, they don't have anything like the taste of wild trout and occasional escapes from fish farms can cause havoc for the native wild species. But it's not the fish's fault – blame the bloke who brought it here in the first place. I think, given the choice, the rainbow trout would have stayed where it was.

Today, there are more than 350 farms in the UK producing around 16,000 tonnes of rainbow trout, 75% of which goes to the table and the rest to restock fishing lakes and streams.

But, oh, if you ever get the chance to taste a bit of wild trout ... it's unbeatable. The brown trout is the only fish I'll occasionally take with me, maybe a couple a year. You drop your fish off at this old smokehouse on the Test – they have a freezer outside and you just leave it in there, so it's a 24-hour, open-all-hours service – and a couple of days later, you pick it up. Once they've been filleted and smoked, they look much smaller, but sweet lord, the taste! There is *nothing* like it.

SIZE

Brown trout grow and develop according to the richness of the water they live in. Some will only grow to about 1lb or half a kilo in a small Welsh or Scottish stream, but they can grow huge given the right conditions. The UK record for a brown trout is 31lb 12oz, caught in the appropriately named Loch Awe.

As returning sea trout have usually eaten considerably more than brown trout that stay in our rivers, there's every chance the biggest brown trout are ones that have returned from the sea – but once they're back in the rivers, their silver colour ebbs away, so they're just brown trout again, albeit bigger.

Wild rainbow trout can reach an average weight of 5lbs, but in the right conditions can be found weighing up to 30lbs. The British record stands at 33lb 4oz.

CHARACTERISTICS

Before we talk about the incredible migrations, let's get two of the best facts out of the way:

- Trout are capable of looking and focusing out of the corner of each eye simultaneously, meaning they have the ability to see in almost every direction at once.
- Trout can vary their colour. Brown trout darken when being aggressive, and when they want to back down, they go lighter. Sea trout also turn silver when they head off to sea, and turn back to the standard brown trout colouring when they've been back in the river for a while.

But there's nothing more thrilling than their remarkable migrations.

All trout are migratory – they need to source food, fatten up and ultimately, return to their home waters to spawn. But one of the varieties migrates much more boldly than the others.

The brown trout stays in freshwater rivers, but through the course of its life, it will move around – from a calm nursery loch as a youngster to the fast-flowing rivers as it grows.

In the case of the sea trout, it migrates from the freshwater rivers of its birth to head out to sea, fill its belly and grow big, before heading back home when it's time to spawn. Their preparation for a voyage at sea starts at about two years of age, when some unknown timer activates inside them and the baby trout go through physiological changes: their skin turns silver, their eyes get bigger, and their internal organs alter to allow them to survive the move from freshwater to salt water. Like the salmon, this makes the sea trout an anadromous fish – one that can live in both types of water.

The sea trout follows the same basic stages as the Atlantic salmon, but once it reaches the sea (travelling downstream at a speed of around 40km a day), it doesn't then keep swimming for hundreds of miles – it tends to feed around the coastline. Studies have shown trout usually stay within 80km of their birth river. Gathering in shoals for the only time in their lives, sea trout might stay out at sea for a few weeks (if the food is plentiful) or over a year. Then they make the journey back, swimming upstream to head home. It's a punishing trip – some will die on the way, and some die as soon as they've spawned, absolutely spent. But many spawn and then, seemingly without a second thought, repeat their journey back to the sea, where they'll fill their faces again and get back to full health. The mass mortality you see with some salmon doesn't happen to anything like the same degree with the sea trout.

WHERE IT'S FOUND

Trout are found all over Britain, but there are lots of areas where they don't appear.

All three species are found in clear waters (which have gravel bottoms, which they need to spawn), be they rivers, lakes or lochs. If they have a preference, wild rainbow trout prefer rivers that are shallow, as well as clear and gravel-bottomed. Having said that, there is only one river in the UK where rainbows have consistently spawned: Bob's favourite river, the Derbyshire Wye.

In rivers, trout can be very territorial over boltholes and feeding runs, but they're less so in lakes where the underwater layout is all much of a muchness.

Since being introduced to the UK in the 1950s by a Danish entrepreneur, farmed trout have been grown in large ponds and tanks fed by a passing river, or with vast net cages hanging in the silent, deep waters of Scotland's lochs. It's a huge business, but it can cause problems for the wild stock.

Farmed trout have been specifically selected for the genetic qualities that are desirable to the fish industry: quick growth and the ability to spawn as early as possible. Which is why it's such a problem when farmed trout get into the river systems.

Farmed trout are domesticated (some farmed trout are the 30th generation of their line to be in farms) and when reintroduced to the wild are completely unsuited to the cut and thrust of nature – they die ten times faster than their wild counterparts. They just don't have the genes or skills they need to thrive in the wild world, but they compete for food with the wild trout, and the inept domesticated genes of any farmed trout lucky enough to pass over a load of unfertilised eggs will dilute the pure native stock.

Just as an example, in 2007, otters chewed through 20 barriers at a fish farm in Perthshire before anyone noticed, and 30,000 farmed rainbow trout escaped. This not only released a load of fish that wouldn't and probably didn't thrive in the wild, but also meant another 30,000 mouths to feed, meaning food was scarcer for wild fish and invariably saw a lot of juvenile salmon and trout devoured by the invaders.

DIET

Larvae, nymphs, crustaceans and insects are the trout's main diet. They live for that magical midsummer moment when the mass hatching of mayflies triggers a mad frenzy of feeding – as do the anglers. It's sometimes called 'Duffer's Fortnight' by anglers, because it's the time when any old hapless cretin will still be able to catch a trout.*

LIFESPAN

Brown trout can live for 20 years. But to get there is a triumph against the odds – 95% of juvenile trout will die before they reach their first birthday.

* Paul: It's not always as easy as they say!

Rainbow trout have a lower life expectancy in the wild, lasting between four and six years.

SCARCITY

Talk to any fisherman over the age of 40 and they'll reminiscence sadly about the years in which they started, when trout (and all fish across the board, to be honest) were more numerous compared to today. This isn't just a rose-tinted view of a halcyon past – it's an absolute fact.

The brown trout is not considered endangered, but it's under a great deal of stress in a lot of places and is labelled a 'priority species' under the UK's Biodiversity Framework, due to many of the outside forces and pressures mentioned above.

In Scotland, the sea trout population has declined greatly, due to sea lice infestations – the most likely culprits seem to be the salmon farms in the sea. Overfishing in the sea can also cut down on the food the sea trout hunts for, and delays their return to the rivers.

In the future there could be additional problems – both trout and salmon require cold running water to spawn, and rising global temperatures could threaten that.

HOW TO CATCH A TROUT

As a game fish, it's often presumed that trout fishing will be expensive, elitist and difficult. It isn't necessarily.

In England and Wales, you need a licence from the Environment Agency to fish for trout, and a full migratory licence if you are fishing for sea trout.

Most game fishing requires you to get right into the water, so you'll need waders that come up over your thighs or chest. For fly fishing you'll need a rod ranging from 7 ft, 3 weight for small streams, to a 9 or 10 ft, 5 or 6 weight for a large, fast-flowing river.

Trout are aggressive feeders and will either be in the mood to take something or they won't. Maggots and worms will catch them easily, but anglers prefer to use the graceful, timeless beauty of the fly.

Some people say that a trout can't determine between a perfect cast and a terrible one – so long as the fly arrives in the water at the moment they're thinking about having a bite. There's also a lot of conjecture that they're affected by barometric pressure, and it encourages them to feed greedily, so the calm before a storm is a great time to catch trout (and plenty of other fish, too).

I've never tried the old poacher's method of tickling trout – once you spot one, you gently slide your hand into the water under an overhanging bank and slowly tickle the trout's belly and, as it relaxes, you grab the bugger – and whether it works, honestly, your guess is as good as mine.

Remember: catch and release applies to wild trout, unless you're fishing somewhere where you're allowed to keep a certain number. If you don't know whether you can keep the trout you catch, then don't!

CHAPTER 10

THE ECOLOGY OF FISHING

Paul: Respect for your quarry is one of the most important aspects of fishing. Ain't that right, Bob?

'To go fishing is the chance to wash one's soul with pure air, with the rush of the brook, or with the shimmer of sun on blue water. It brings meekness and inspiration from the decency of nature, charity toward tackle-makers, patience toward fish, a mockery of profits and egos, a quieting of hate, a rejoicing that you do not have to decide a darned thing until next week.'

Herbert Hoover

Paul: I spent most of the time until I was in my twenties bait fishing, coarse fishing, all that freeline stuff. And you know what? I was a much better fisherman when I was in my mid-teens.

Bob: Than you are now?

P: Oh yeah. As a teenager, you try everything, you absorb everything. You never grumble and give up and go home. You never even consider it. So you can't fail but to do well, because if it's not happening, you're going to try this, then you're going to try that … endless enthusiasm. Back then, if there were chub under the trees, I'd go out and I'd get one. I'd *definitely* get one. I just had that attitude.

B: But there were also more fish back then, weren't there, Paul?

P: Certainly were.

B: Lots more back when you were a teenager. In the mid- to late-1920s, yeah?

PAUL

When we first thought about making *Gone Fishing*, Bob did have a slight concern that fishing was not politically correct these days. He's got a point. I think about it a lot, and I know it's something that really unsettles Bob.

Paul McCartney, who might be the world's most famous vegetarian, did a campaign a few years back about going veggie and he mentioned he'd given up rod fishing. He was out fishing many years back and it struck him that this struggling fish on the end of his line wanted to live – and he was killing it for the fleeting pleasure the sport of it brought him. The fish's life was as precious to it as McCartney's was to him. He gave up fishing and eating fish, and hats off to him.

I have a huge amount of affection for the fish I catch. It might be one way and they might really resent meeting me, but I love meeting them. And most of the fishermen I've met in my life also have an utter love for their quarry. They feel so strongly about fish to the point that I think a lot of them – not all, but a lot, a significant proportion – would be happy to stop fishing immediately if it was doing significant harm to those fish.

I've seen it happen. When stocks are low in certain rivers and the Environment Agency says, 'Right, no more fishing this river,' the fishermen not only accept it without question and move on, but most of them will be saying, 'About time, I've been saying they should have taken this measure for years.'

Take catch and release. Today, all coarse fish are released back into the river after they're caught. This is to ensure that fish numbers are protected. Catch and release has been a standard practice for most people and has been a part of the culture of fishing in the UK for over a century. It's regarded as a cardinal sin and a capital offence not to return the fish. Eternal damnation. This is certainly the case with coarse fishing. Not so much with game fish, but times are changing.

For people of my dad's generation, fishing was driven far more by economic necessity. Coming from the deprived hillsides and valleys of Wales in the 1920s, if you caught a trout, you wouldn't have put it back. That fish might have been one of the few things that stood between you and hunger for the next couple of days. Some might argue that this is a purer form of fishing – you're catching it to eat, out of necessity, not catching it for sport and toying with it for the enjoyment that brings. But most fishermen today have two loves: a love of fishing, which is something we've done since we started walking upright, and a love of fish, whom we respect enough to say, 'We don't need to eat you any more, so we're going to put you back, and God bless.'

It's a strange urge, isn't it? You want to both celebrate the fish *and* capture them. It's part of human nature – think of the native tribes who go hunting. They're going out to capture and kill their quarry, but they do so with a total and utter respect for it. Most fishermen are the same, except they don't kill. And anglers do more, by far, than all other conservation groups to ensure survival of fish and improve water quality.

That moment when you put a fish you've caught back into the river, and they take a moment to get their bearings – I can't help but feel there's a brief second where they realise they're being freed and are genuinely a bit shocked and pleased about it – and then with a flick of their tail, they're away ... That's lovely. It's as good as the fishing itself.

If I go fishing now, I would find it hard to kill a fish. As I previously mentioned, if I'm fishing in Hampshire, I do occasionally take a trout with me and take it down the road to a little smokery and have it smoked. If I'm going to eat fish, which I do, then I reckon I should be aware of what it entails and be prepared to do the dirty work myself, rather than delegate it to someone else.

Some of you will say, 'All right, then, Paul, if that's how you feel, you should also be going down to the slaughterhouse and

Paul: A beautiful Icelandic trout gracefully demonstrates catch and release.

murdering your own cows,' but I don't tend to eat meat any more, so pick the bones out of that.

When catch and release was introduced almost across the board with salmon, there was a certain amount of backlash. Game fishers felt that catching salmon was something that had been going on for thousands of years – for as long as there's been fishing, for as long as there have been people on earth.

Plus, these are fish that we eat and buy in the fishmongers', so why can we take them from the shelves of the supermarket – and God knows how they've been caught, but I can tell you it's a much more unpleasant way than on the end of a rod – and not from the river?

I'm not going to get into an attack on commercial fishing here, but if we're looking for a reason that salmon stocks have declined, and an industry that seems to have little serious regard for fish welfare, then it's probably worth having a long, hard look at them before anyone else. Though it is important to acknowledge that some traditional commercial fishing has maintained communities for centuries and the fishers have lived in harmony with the fish. Historically, there were far greater fish stocks and no industrial trawlers capable of hoovering up a whole run of salmon destined for one particular river, for example. Times have changed so we must too, in order to preserve and conserve stocks.

If you're a fisherman taking a single salmon (which of course you'll check to make sure it isn't carrying eggs), then you dispatch it quickly and humanely with a small bat (it's also known as a priest, because it administers the Last Rites). It takes a few seconds. Commercial fishing vessels just haul hundreds of fish out of the sea in nets, tip them in the hold and leave them to flap around until they finally suffocate.

So to many game fishermen, releasing a wild salmon felt like a hard pill to swallow – until the first time you put a salmon back. It's both very difficult to put a salmon you've finally managed to catch back in the river, and an utterly amazing thing to do. It's such an extraordinary creature. Its wildness, the colossal journey it's undertaken, its iconic position in fishing lore – if I call a roach life-affirming, then a salmon makes you want to praise the universe. Not to undermine any other fish, but when you have that salmon in your hands and it suddenly gives a kick of its tail, and then it's gone – that's great. In fact, I can't actually imagine not releasing a salmon any more.

But you have to admit, the catch and release programme hasn't saved the salmon. Since the early 1970s, the number of salmon in England and Wales has declined by 50%. Numbers were terrible before, but now it's gone off the scale. In 2014, wild salmon stocks were at their lowest level ever recorded, and they've continued to do badly ever since.

This isn't just frightening news for the salmon – it's very worrying for a lot of people, too. Salmon fishing in England and Wales supports 900 full-time jobs and provides £22 million of household income, mainly to the rural communities and economies where salmon fisheries are located. If there's no salmon fishing, there's no income.

And think of how this unchecked decline will affect Scotland. Take the Spey. It was the holy grail of salmon fishing – probably the best out of the big four on the east coast. People would come from all over the world to fish it. For years, you could not get on the Spey for love nor money: it was dead men's shoes. Nowadays, you can get on it easily. That's because the salmon's declined so much. People aren't going to pay massive sums of money to go and fish, book a hotel, use the restaurants and pubs, and pop into the tackle shops if there's almost no chance of catching a salmon. The economies of those remote parts of Scotland have been kept alive for countless years by anglers, but they depend on the salmon being there to catch.

At the times of year when nobody would be going to Scotland beyond a handful of skiers, fishers would be all over the rural areas. I've fished the River Findhorn in blizzards in February and March, but who is going to go if the fish aren't there? It's such a concern. Not just for their tourist industry, for everyone – the salmon is a species that's synonymous with Scotland and it's just disappearing.

The problem with dwindling salmon stocks isn't the anglers. It never has been. But no one can say definitively what the problem is.

What is known (from a large report commissioned by the government's Environment Agency in 2018) is that there are far fewer young salmon, at every stage of their development, in the 49 rivers in England and the 31 rivers in Wales that support salmon.

Most salmon populations have declined, in some cases severely, and are not predicted to improve in the next five years. The numbers of rod-caught salmon are currently amongst the lowest ever recorded. Around 70% of juvenile salmon heading for the sea for the first time die before reaching the coast. And only a tiny proportion – something like 5% – of salmon who successfully make it out to sea will ever return. This fish is under so much pressure. And there's a combination of factors that are causing this.

Commercial fishing is one. It's not the only pressure and the Environment Agency have said it might not even be the main cause of dwindling stocks in our rivers, but it's certainly a factor in the decline of the salmon.

It's joined by the spread of salmon lice. These lice attach to a passing salmon and then begin eating them alive, leaving great, ragged wounds on the salmon's body that are open to infection. The Atlantic Salmon Trust cite considerable evidence of a link between intensive salmon farming off the coast of Scotland and the spread of these lice to both wild salmon and sea trout.

For a brief time, these fish farms looked like they were going to be the saviour of our native wild stocks. The thinking was that with such cheap and accessible salmon, people wouldn't need to catch the wild ones in huge nets any more. But it's not worked out like that. It might seem like a good employer for a rural area, but it's not. I'm happy to say it: it's a dirty industry that doesn't employ many people, and the practices many of them use are appalling.

Packing thousands of farmed fish closely together means an ideal breeding ground for the lice, and so they pour chemicals in to try and reduce the infestation. But the farmers are playing catch-up and they're a long way behind. The lice have already developed resistance to a number of pesticides, and the effect they're having on the wild fish (who let me remind you don't, by and large, have access to sea-lice pesticides) is irreversible and devastating. The sea trout has already been practically wiped out on the west coast of Scotland as a direct result of the lice from salmon farms. Maybe in a few years we'll be saying the same about the salmon.

Pollution of our rivers is still a major problem. The old days of entirely black canals and waterways, the water and everything in it poisoned by the off-swill of heavy industry, might have ebbed away but many of our rivers are still being polluted. There's an awful lot of agricultural pollution, which is less obvious, less tangible and less visible, but does untold damage.

You might walk down a river in the countryside and think, 'Look, there's this lovely unspoiled river running through a sylvan valley, with a lovely old pig farm just sitting back on the brow of the hill there – what a perfect scene.' But what you can't see is the phosphates from the fertiliser and the slurry used on the farm leeching into the river systems every time it rains. It's a silent killer.

In 2017, the Environment Agency said there was a 'serious pollution incident' every single week in the UK, where slurry from livestock farms ended up in a waterway. A 2016 report into Wales's freshwater habitats found that just one in six were considered 'favourable' for wildlife. In 2018, the Rural Affairs Secretary Lesley Griffiths told the BBC that the scale of agricultural pollution in Welsh rivers was 'embarrassing' and promised tougher regulations would be brought in by 2020.

Our rising population also puts pressure on our rivers. Abstraction – where they take water out of the rivers for man's use – is also a massive problem. Every time I hear people saying, 'We're going to build more homes,' I think, 'Oh, you poor bloody rivers. That's going to mean more abstraction, more pollution ...'

You can add to the pressures facing fish the risk of being predated by other animals. Young salmon are eaten in huge numbers by birds, but it seems no one's willing to look at schemes to try and protect the fish.

Cormorants are one of the worst offenders. These large black birds are known in fishing circles as 'the Black Death' because of the numbers of fish they consume. Anglers on the Walthamstow Marshes report they see some 200 to 300 cormorants flying over in the course of a day's fishing.

For two decades, 11 groups, including the Angling Trust and Salmon & Trout Conservation, have been lobbying the government to review the problems caused to fish numbers by cormorants. It's never been looked at. The RSPB's senior policy officer told the *Independent* that anglers must find ways to live alongside them'. It's not the anglers who need to find ways to live when the cormorants are around, it's the fish.

In 2018, the Dee District Salmon Fishery Board asked for 'a sensible debate' about a cull of birds like the goosander duck, after they found that 70% of their young salmon weren't surviving long enough to make it to sea. The RSPB responded by saying that birds were only one of the pressures on the young fish, and 'we would be very surprised indeed if the authorities decided to increase the killing of native wild predators on the basis of such slim new evidence'.

And these are just the problems for the salmon in our rivers. It doesn't take into account everything that can happen to the salmon while it's at sea. On the way there: predation, pollution, industrial fishing. On the way back: predation, pollution, industrial fishing. Seals, for example, take countless salmon, but who's going to say, 'Oh, we should get rid of a load of those seals.' No one! People love seals, with their fluffy coats and their big emotional eyes. It's the same with dolphins. 'Oooh, look, lovely dolphins!' Well, the salmon isn't thinking, 'Lovely dolphins.' He's rolling his eyes and saying, 'F*ck me, bloody dolphins, seals, cormorants …', although chances are he won't be thinking it for long before he gets tangled up in some industrial fisherman's net.

And as if that wasn't enough, all the above could be made moot as the method by which the salmon will become extinct, because the rise in global temperatures might be able to achieve it in the space of just a few years. Many fish, like the salmon, require cold water to spawn in and even a small rise in this water temperature will see their eggs rendered useless. A couple of degrees up and the salmon is over. Finished. For good.

Amazing, isn't it? It's 2019 and the salmon, a species our ecologically aware nation both admires and has a better understanding of than ever before, is somehow in the worst place it's been in for 60 million years. We're seeing our native salmon die out in front of our eyes.

And it's not just the salmon and the sea trout. All of our native species are struggling. The roach has declined from the Avon and the Wensum *catastrophically*. It's very difficult not to look back through rose-tinted spectacles to your childhood, but fish were certainly more plentiful back then. Talk to any fisherman over 50 and they'll agree.

What underlies all these declines is the common denominator: our rivers. Salmon are an iconic indicator of healthy river systems, so if the salmon is doing badly, it means our rivers are, too.

John Bailey, the UK's best-known angler, told me he thinks that the last 25 years have seen our nation's rivers more seriously damaged than at any other point in history. And bear in mind, during the Victorian era, a lot of our urban rivers were dog graveyards made of tar and human shit.

The odd thing about fishing as a hobby is that when you start, you want to catch fish. So all fishermen want more fish in the rivers. And yes, they want clean rivers. And yes, they want well-maintained ecosystems. All these things make fishing easier and more enjoyable. But it doesn't take long for a fisherman to reach a point where they don't just want those things for themselves – they want them so that future generations will be able to get the same pleasure that they have, and they want these things for the fish.

Fishermen become deeply attached to the rivers they fish. It becomes a sacred place. It's a haven and a retreat. They know the contours of the bank like the face of an old friend. They have their favourite trees, they know the point to see to get the most out of the setting sun, and they can walk every single inch of it in their dreams at night. And what happens when you have a place like that is you want to protect it.

One of the problems we have in the UK is that the two main players involved in river management tend to be at loggerheads. On one hand, you've got the academic players – the marine biologists and environmentalists – who want to see habitat improvement done naturally, for fish stocks to return of their own accord, and to leave these areas as natural as possible. And on the other hand, you have the anglers and the people who work on the rivers. They want all those things as well, but they also want to be able to have fish in the rivers so the businesses that rely on them can keep going.

There's a big battle here between the people working on the coalface of the rivers – the gillies, the guides and the anglers – who want to stock rivers with, say, salmon from the existing brood stock. The genetic integrity of salmon is crucial because, it's argued, they've got to return one day to this same river. But the environmental and academic groups, who seem to be in control of the overall river system, will say, 'No, we're not going to stock the river. If we make it right, they will come back.' But they're not going to come. They aren't there.

There has to be a compromise between wanting fish to be living in our rivers and wanting rivers made absolutely perfect for fish. Because unless there's fish in the river, you can have the best spawning area you like, but if there's no fish around to use it, it's pointless. There's nobody home.

The anglers argue that our rivers aren't untouched natural habitats – they've always been managed to one degree or another. Take the Test – some people call it the longest stock pond in the world. The chalk streams across Britain are nothing if not a managed environment, and they have been for centuries – like most of our countryside. Even the wild ones that we think just *exist,* they don't. They do not just exist.

Fishing rights on most of the rivers in this country are split up into small stretches, with the riverbed technically owned, controlled and looked after by an individual owner. While some are private individuals, many are owned by trusts and fishing clubs. The money these owners receive from issuing permits to anglers usually goes straight back into the upkeep of their stretch of river. And a lot of work goes into preserving and protecting our rivers. If these rivers are in a good natural state – or like the Thames, the Tees and the Tyne, rivers that were formerly heavily polluted that have come back to sparkling, glittering life – it's only because people have worked really hard to get them there.

In a way, when it comes to the survival and improvement of the habitat and fish species, I'm sure a lot of the eco-groups begrudgingly admire the fishers. There have been a lot of angling groups, along with the Angling Trust, who've taken polluters of rivers to court, and anglers are conscientious reporters of strange sights and invasive species. And fishermen are the guys on the ground. They're the watchdogs, looking over and fiercely protective of the fish and the river habitat.

The key thing is fishermen know these places – their route, their particular beat – better than anyone else. They're down there more often than anyone else, which means they're the best-placed sources to note any changes. They notice if the water is changing its rate of flow. They can see changes in the surrounding environment. And they're aware when the fish numbers are declining. But they're often the last people to be listened to, because their knowledge doesn't come in a folder with a load of footnotes, and they're seen as being self-interested because there's a rod in their hand.

We know the rivers and the fish have had it hard in the past. They're having it hard again. I just hope we can halt some of the pressures on them. Many of the things that we allow to happen are much, much crueller than fishing. Fishermen might have a self-interest but they're also always fighting the fish's corner.

We've saved our rivers before, and I hope we'll do it again. Maybe there's hope. Maybe.

CHAPTER 11
ON RIVERS

Bob: The locations are one of the best things about our fishing trips.

> 'The traveller fancies he has seen the country. So he has, the outside of it at least; but the angler only sees the inside. The angler only is brought close, face to face with the flower and bird and insect life of the rich riverbanks, the only part of the landscape where the hand of man has never interfered.'
>
> *Charles Kingsley*

Paul: You like the rivers we go to as much as the fishing, don't you, Bob? If not more.

Bob: I do, yes, Paul. Do you remember when we were on the Wensum? That was lovely. And we caught roach after roach after roach.

P: And then there was that day on the Test when I went off to catch pike on the fly while you trotted off for grayling. Something I'd not done much of, but Christ! I landed a big lady with that.

B: Had a good old scrap with her, did you?

P: She went off like a shot and then through that little jetty – oh, that was fun.

B: That was a proper fishing chat, there, Paul. I enjoyed that.

BOB

I like fishing a great deal, but I like the places that fishing takes me to even more.

People have different aesthetics, but I think it's pretty hard to beat the environment of a river. And these beautiful places are right there on your doorstep, that's the wonder of it.

If you asked the question, 'Why do people fish?' then I'd suggest you look at the lovely episode of *Gone Fishing* we did where we're fishing for sea trout on a little pontoon on a beautiful summer morning. I know you'd watch that and you'd think to yourself, 'I wish I was there.' That's nothing to do with me and Paul, or even the fish – you'd think to yourself, 'Oh, I'd like to be standing there in the sun in Christchurch.' It's magical.

If I was a collector and fishing was my full-time hobby, I'd be more interested in ticking off rivers than ticking off species or sizes of fish. I'd be, 'Right, I'd like to go to the Tees, and the best bit of the Usk, or the nicest part of the Stour.'

It'd be nice to go salmon fishing on the Tyne. I knew the Tyne from my childhood and I was brought up thinking the Tyne and Tees were barren. For many years, they were. But the Tyne's now doing really well, the salmon have returned, and it's one of the most celebrated regenerations of a river in modern times.

I think my favourite river is the Herefordshire Wye. I'd happily fish on the Wye and nowhere else. I bought a book from the 1920s all about the Wye by a man called AC Bradley, but it turned out to be the single most exhausting book I've ever read. I don't know if AC Bradley was being paid by the word, but if he was, he must have ended up the richest man in Europe. Here's a sample:

> As if the Wye, blessed above all rivers and through all its days, had not enough of fame, it breaks into the world, a tiny rill, in such a fashion as to set all the bards singing, and even to fix the story of its birth on the reluctant mind of countless schoolboys. Strangers not a few make the long and trackless pilgrimage to the wild glens where, almost within shouting distance of one another, the triple streams leaving the bosom of this great mother of waters, head for such different worlds.

You can boil all this down to the sentence: 'The Wye has a source.' I thought it might have some tips on where the nicest bits were, but it doesn't, it's just 189 pages full of what Russell Brand would sound like if he was an Edwardian. But it's a reminder that it's actually quite hard to write about a river. It is, after all, just some water in a mud chute. Rivers were made to be stood in front of and admired with the eyes and absorbed by the soul. You can't do them justice with words. So don't complain this chapter is just a list.

BOB'S TOP FIVE UK RIVERS

1. THE WYE

The fifth longest river in the UK, the Wye runs for 134 miles from mid-Wales to the Severn Estuary, and forms a lot of the border between England and Wales. It flows through some of the most beautiful towns you'll ever see – Hay-on-Wye, Hereford, Ross-on-Wye, Monmouth and Tintern. Even the poet William Wordsworth thought the Wye was ace:

> How oft, in spirit, have I turned to thee,
> O sylvan Wye! thou wanderer thro' the woods,
> How often has my spirit turned to thee!

The crystal-clear waters of the lower Wye mean it's one of the best rivers to fish for salmon outside Scotland. It's also good for trout, grayling, barbel, dace and chub.

2. THE TYNE

Full disclosure, I've not fished the Tyne yet. But I fully intend to. Running from Cumbria to the North Sea at Tynemouth, the 73 miles of the Tyne was for many years lined with shipbuilding yards and ships exporting coal, the epitome of an industrial river with all the pollution that came with it.

With the decline of that industry, the Tyne has come back to life, and it's now the best river in England for salmon and sea trout fishing, bucking the trend that's seen numbers of these fish decline all over the UK. You can watch them jumping the weir at Hexham as they return from the North Sea to their spawning grounds. If you'd told me that as a kid, I'd never have believed it. I'd have thought you were trying to trick me, or make me look stupid, or abduct me, and I'd have run away.

Coarse fishing includes dace, roach, chub and gudgeon.

3. THE SPEY

I'm not there with the fly fishing yet, but this is an iconic river that if you're a fisherman, you dream of fishing on. It's the fastest-flowing river in the UK, 107 miles of water running from the Highlands into the Moray Firth and famed for its salmon and sea trout fishing over the last century.

The origins of the river's name are disputed, but one theory is it's derived from the pre-Celtic *squeas* meaning 'vomit'.

Part of the joy of Scottish rivers is they've been doing this for years, so the hotels are some of the best you'll ever stay in. Speyside also has a lot of whisky distilleries, producing more whisky than any other region, so if the weather's bad, you can get really drunk instead. Or if the weather's good, it's your holiday and you can spend it as you see fit.

4. THE TEST

I've spent a lot of time fishing the Test with Paul and it's a cracking river. A chalk stream running just 40 miles from Ashe in Hampshire to Southampton, it's the sort of idyllic southern English river you'd see in an old episode of *Poirot*, where they go out on a punt and an art deco woman is strangled by an adulterous maid.

It also features in *Watership Down*, the best animated film ever made about rabbits who start murdering each other in a cold-blooded power struggle.

The birthplace of the dry fly, the river is stocked with salmon, trout and grayling. Coarse fish are pike, perch, roach and dace. It's expensive to fish there, but it's a lot cheaper in the winter when I should imagine it's just as beautiful, if not more so.

5. THE THAMES

Second only in length to the Severn, the 215-mile long Thames is another industrial river that's come back from the dead and is now home to a huge variety of some 125 species of fish. In the 1960s, it was declared 'biologically dead' and if anyone fell in, they were genuinely put in quarantine.

It starts in Gloucestershire and runs all the way to the Thames mouth in Southend-on-Sea, which has a great little railway on its pier if you're ever there and want something to do. It runs through Oxford, a stretch which is called the Isis (don't join a fishing club there – if you are on holiday in the USA and they find the membership card in your wallet, they'll fly you to Guantanamo Bay) and runs right through London, where it shakes its head sadly at the house prices.

Today, the Thames is home to bream, perch, pike, roach, rudd, dace, barbel, carp, chub and gudgeon. And as a river running right through a major city, the Thames also offers fishermen the added excitement of accidentally catching handguns, syringes and the corpses of East End villains who grassed.

CHAPTER 12
FLY FISHING

Paul: Ted and Ralph enjoy a spot of fishing.

'Angling may be said to be so like the mathematics, that it can never be fully learnt; at least not so fully, but that there still be more new experiments left for the trial of other men that succeed us.'

Izaak Walton, The Compleat Angler

Paul: I've always gone fly fishing, but I've been doing a lot more float and ledger fishing since we've been doing the show. The TV programme's reminded me just how much I enjoy it.

Bob: I am also going float and ledger fishing a lot more, given that I never actually went fishing at all until now.

PAUL

Since 1872, the firm of Hardy Brothers have been supplying fishing equipment to the great and good, and in their catalogue of 1907, they summed up the basic principles of fly fishing better than anything ever written since:

> The art of dry-fly fishing is to present a fly that floats – and floats perfectly – to the notice of a rising fish in such a manner that it is mistaken for the natural *ephemera* which is hatching out, and in the result is accepted as such by the fish. To lure *Salmo Fario* successfully after this manner it is necessary that the angler should have skill; be very observant; have the patience of Job, and, beyond all, be properly equipped for the task.

Yes, they slip in some unnecessary Latin because they're posh, and the last line is a shameful attempt to flog you one of their rods, but we can forgive them that.

I started doing a bit of fly fishing in my late teens, maybe 20s, and without question, fly fishing is widely seen as the most elite form of angling. It isn't, but historically, it was seen as being so pure and unspoiled that it made all other methods of fishing seem positively coarse (which is exactly why it's termed 'coarse fishing'. Clearly, when they had the meeting to decide what the various forms of fishing should be called, the roach and tench fishermen arrived after the final vote had been cast).

Before I go any further, I should probably tell you where I stand on this thorny issue: *I don't really pay it any attention and I certainly don't want to make it worse.* Look, coarse and fly fishing have a lot of similarities and some notable differences, but fishing is a broad church and it's silly to argue that one method is inherently superior to any other. So long as the angler is enjoying him or herself and the fish are treated with respect, then that's fishing at its best.

When I go out, I do prefer the fly, but in recent years, I've re-discovered the joy of trotting a float – something I did with my dad and my mates growing up. I'd wager that a lot of the people who do fly fishing started as coarse fishers. And if you asked me to go out for a day doing either, I'd bite your hand off, whether we were going for salmon or perch. It's horses for courses.

And let me make it clear, the absolute last thing I want to do is in any way irritate or put down carp fishermen. After all, I've seen the terrifying hardware they employ to catch carp – a quarry they love and respect. And I don't want to think about the hi-tech weaponry they'd use against me if I upset them and they turned on me. I wouldn't stand a chance! If I upset the carp fishing community, you'll know, because I'll be found lying dead on a canal towpath or in the bushes at a commercial fishery within 24 hours of this book hitting the shelves.

The two main branches of freshwater fishing in the UK are defined by the spawning times of the species involved and also some colossal historical snobbery based on the class system we all enjoy in these islands.

Actually, the British class system has been good for two things: humour – oh, how we laugh at the upper, middle and working classes – and some exquisite country houses and estates that the 'poor' aristocracy now have to open to Joe Public. We can snoop around their gaffs to our heart's content before pondering how emotionally fragile the nobility have become; it's obvious to them now that they're no better than the rest of us. Rant over.

But this mob in their Victorian heyday were the ones responsible for defining freshwater fishing as either game (salmon, sea trout, trout) or coarse (barbel, carp, roach, perch, chub, pike … In fact, all the other species that swim, including grayling). However, grayling are a slight anomaly in that they're often fished for with a fly or nymphs but live in the same rivers and conditions that trout and salmon like. They also share an adipose fin with trout and salmon but have their oats at the same time as the roach and tench. I've confused myself now! This makes grayling difficult to categorise.

As you might have gathered, I love all branches of angling and certainly don't see one method as inherently superior to another. But I enjoy fly fishing the most, closely followed by trotting a float or watching that float against a backdrop of lilies and fizzing tench bubbles.

Fly fishing appeals to me because I don't have to cart loads of gear or bait around with me. A day fly fishing on a river requires little more than a nine-foot fly rod, a reel and line, a box of flies and various small accessories, snips, leader material, floatant and so on, all of which can fit into a pocket or two in a fishing vest. You can be ultra mobile … that sounds a bit too dynamic. You can wander freely up and down the river, enjoying the timeless wonder of the English countryside, the wildlife, the roar of fighter jets as they scream overhead, shattering the tranquillity as you search for an interested, rising fish. Even salmon fly-fishing tackle can be simplified to suit the prevailing conditions and allow you to roam free in your mighty chest waders. If there's not much happening on the angling front, take time to study nature, think about your loved ones, maybe compose some poetry or contemplate just how big a halfwit Bob Mortimer is.

Like the hunter, the fly fisherman stalks his prey, actively tracking it down, getting eyes on it, attempting to second-guess its next move, before taking that one single shot with a rod, which, if accurate, will bag him his prey. As such, fly fishermen often unfairly portray bait fishermen as mere trappers, lurking in a shrub in the drizzle, spending the empty, endless hours waiting for a bite. Unlike the debonair fly-fishing gentleman, who strides around the roaring river in full flow, picking his shots with dazzling accuracy and pulling out game fish the shape, colour and value of bars of silver.

This is, of course, arrant nonsense. Whatever your quarry and whatever your method as a fisher, you have to become immersed in the environment and be stealthy rather than invade it; unless you're a carp fisher, because a lot of them do look like they're going to Helmand Province.

It's pointless to claim one form of angling is better than another. I mean, if you boil it down to its bare bones, fly fishing is just the act of conning a fish with a home-made insect, which barely resembles the actual insect we're trying to imitate. Hardly the pinnacle of human artistic achievement, but fly fishing for our usual quarry – salmon, trout, sea trout and grayling – often takes place in some of the more rural, untamed and wild places. Some places of staggeringly raw beauty and some historically managed environments with extraordinary water quality and purity.

If I had to pick one form of fishing it would be for salmon with a fly, or more accurately, a lure, tied to represent a small fish or a shrimp on which salmon feed at sea. Salmon are truly wild creatures that are often hard to catch. That's an understatement because the buggers don't actually eat *anything at all* when they're in our rivers, so catching one with a tasty-looking morsel seems to be the triumph of a fisherman against not just a fish, but logic itself.

And once hooked, these lads fight to escape with power and passion. These fish have effectively evolved with two distinct urges: (i) to get back to the rivers they were born in so they can spawn, and (ii) not to get killed before they do. Many of these fish will have spent months travelling thousands of miles through the vast Wild West of the open sea, avoiding predators on a constant basis, always paranoid, always wary, always suspicious.

Amongst the obstacles to their extraordinary journey a thousand or so miles back from their feeding grounds around Greenland are any number of predators, including dolphins, seals, industrial fishing vessels, more seals, otters, netsmen and Whitehouse (the latter being the least successful of the salmon's natural predators).

The take of a salmon to a fly is something so magical that it's impossible to describe so I'm not even going to try. But once you have them on the end of a line, that's not the end of the matter – not by a long chalk. These fish EXPLODE and will do anything and everything imaginable in their power to escape. It's amazing to witness their sheer iron will to live and their ceaseless determination to be free. Honestly, the moment you catch a wild salmon or a big sea trout on fly tackle, you have never seen so much life being lived so vigorously in such a small amount of space.

With this in mind, losing a fish is nothing to be ashamed of or to be irritated about – it's all part of the rough and tumble of the sport. So the fish managed to win that round – 'Well played, fish, you were a most worthy opponent who got the better of me this time!' – and you dust yourself off, straighten your rod, and get straight back into the game for round two. Let me just say here though that doesn't mean you should be in any way proud if a fish breaks your line. That is just not on. It means you haven't prepared properly or are using inappropriate tackle. Unless there are extenuating circumstances, I will shun you!

The art of fly fishing cannot be mastered. Even the most experienced fly fisher, with a lifetime of practice and a casting arm that can castrate a passing bluebottle in the blink of an eye, will be outfoxed by many, if not most, of the fish he comes up against.

But if the fly fisherman is consistently successful, it's nothing to do with luck – it's the direct result of their command of the art. Success in fly fishing depends on the countless tiny decisions the fisherman makes before the line is even cast, and subsequently, the skill to execute those decisions exactly.

Or if there are no fish rising, the fly fisher has to cast the fly into likely-looking spots, or use a sub-surface nymph and try to induce a take from a fish by imparting movement to duplicate the action of the insect he's copying.

And that's just the mechanics of the fishing. Fly fishermen don't only need to be able to cast to the precise spot they choose, but understand *why* that exact spot is the right one. Once they've spotted a rising fish, they have to try and decipher their meaning – has the fish risen and dropped back down, but remained in place? Or did it pop up just as it was swimming away? If it did swim away, how far might it have gone? In which direction? Two feet downstream? Or ten feet towards the bank? Has it taken off entirely? Even if he can see the fish through the water, where should he place the fly to have the best chance of it being taken – directly above the fish? Or a foot away so it drifts down on the current? Should he move it to mimic a real fly? Or will that spook the fish?

The answer to all these questions is – there is no answer. And frankly, it's doing my nut in! But you need to decide swiftly, or the moment will have passed. Experience is the only thing that can guide you through these endless questions, and even then, there are no guarantees.

If there were, fishing would lose its mystery and appeal. There is an old adage about fishing hell being to see a fish rise, casting to it and catching it. Seeing a fish rise in the same spot, casting to it and catching it, seeing another fish rise, casting to it and … you get the picture … the same fish from the same place ad infinitum, ad nauseam. I think that's the correct Latin but we didn't do Latin at my school. Ask Bob, he trained to be a lawyer.

So it's about making quick decisions that come with experience and practice and if fortune isn't against you, you'll have taken on nature and won.

Paul: Me and the fish from the river Dee.

I remember I was once salmon fishing on the River Dee, where I go quite a lot. I saw a fish move quite a long way out, so I got into the river, waded out and I cast a specialist cast called a single spey cast. The fly swung round sweetly and the line drew out magically as the fish took straightaway and I landed it. I was only able to catch that fish because I've been doing it for a long time, so I instinctively knew what I had to do. I knew I'd have to wade to a certain point, I knew exactly where the fly needed to land above the fish, I knew the only cast that would get out there was quite a specialist cast, and I knew the fish was going to take because of the way it moved. I wasn't consciously thinking of every stage – I just knew what I had to do, and I knew every step I had to make to make it happen. And it did.

Mind you, let's get it right: there are loads of times I've done the same thing with absolutely no result!

But I remember that fish so well, because as I caught it I thought, a lot of experience and accumulated knowledge over the years has gone into catching that fish. It wasn't a particularly big fish at all – it was just a coming together of circumstances that made it very memorable. That was a terrific sense of achievement.

To head out armed only with a thin rod, a line, a handful of fluff and hard-won years of experience, and end up catching a beautiful wild fish – to me, that's angling in its simplest, purest, most thrilling form.

There's one other big bonus about fly fishing: it liberates you from the tyranny of tackle. No more maggots in a bucket, stinking of ammonia and inevitably getting loose in the car. You only discover some of them escaped a week or so later, when you're about to set off to the supermarket on a warm summer morning and open the door of the Auris to find it packed full with so many bluebottles, it's like you're in a deleted scene from *The Amityville Horror.*

When I was just starting to go salmon fishing, I knew this fishing tackle dealer from Birmingham. He said to me one day, really sadly, 'Listen, Paul, I'm fed up of the maggot game. I don't want to spend my life scrabbling about with maggots.' And I understood exactly what he meant. In the words of FH Halford:

> ...the cleanest and most elegant and gentlemanly of all the methods of capturing [fish]. The angler who practises it is

> saved the trouble of working with worms, of catching, of keeping alive, and salting minnows, or searching the river's banks for the natural insect. Armed with a light single-handed rod and a few flies, he may wander from county to county, and kill trout wherever they are to be found.

It's hard to disagree with any of that. Apart from the bit at the end where he suggests killing as many trout as you can wherever you can find them. However, since I started trotting for grayling again about seven or eight years ago – and especially since fishing with Bob and the great John Bailey on our series – I've fallen back in love with the maggot. It's like we've never been apart. Be still my beating heart.

THE HISTORY OF FLY FISHING

Fishing with a rod, line and an artificial lure created to imitate the prey of fish – those three essential ingredients of fly fishing – was first recorded in the year AD 200, when the Roman author Claudius Aelianus watched some Macedonians fishing in the Astraeus River:

> … they have planned a snare for the fish, and get the better of them by their fisherman's craft… They fasten red wool…round a hook, and fit on to the wool two feathers which grow under a cock's wattles, and which in colour are like wax. Their rod is six feet long, and their line is the same

> length. Then they throw their snare, and the fish, attracted and maddened by the colour, comes straight at it, thinking from the pretty sight to gain a dainty mouthful; when, however, it opens its jaws, it is caught by the hook, and enjoys a bitter repast, a captive.

That said, Claudius might not be the most reliable source: in his *De Natura Animalium* (*On the Nature of Animals*), he claimed that beavers were famous for gnawing off their own testicles to throw into the path of pursuing hunters, so best to take everything he says with a pinch of salt.

While it's most closely associated with the Victorian era, fly fishing had been practised in England for centuries – Dame Juliana Berners and Izaak Walton's books both have considerable sections about making artificial flies.

Alfred Ronalds' *The Fly-fisher's Entomology* (1836) was the first book to start giving names to individual artificial flies and detailed instructions on the time of year to use them.

An engraver by trade, the pages were full of beautiful, accurately realised illustrations, and his scholarly book helped to foster the serious and academic air that the majority of books on fishing still employ today (this one, I'm aware, is the exception that proves the rule). For better or worse, Ronald was the man who laid the groundwork to turn the act of simple fishing into a complex and gravely serious science.

The success of the book led to the longest lasting legacy that the Victorians left to fishing: the development of the art of the artificial fly. The creation of little handmade fake flies – which were designed to trick a particular fish on a particular day, often at a particular hour – with just a handful of feathers and thread distilled the main four Victorian passions into one respected pastime. It combined engineering, scientific learning, technical ability, and it kept your hands occupied for hours, so you wouldn't be tempted by the dark delights of sexual self-pleasure.

The art of making your own flies – fly-tying – became more than a hobby during the 1800s: it was a mania. A lot of factors contributed to this perfect storm: the Victorians loved to classify and catalogue anything and everything. Crafts and handiwork were universal pastimes. The natural world was in vogue (let's face

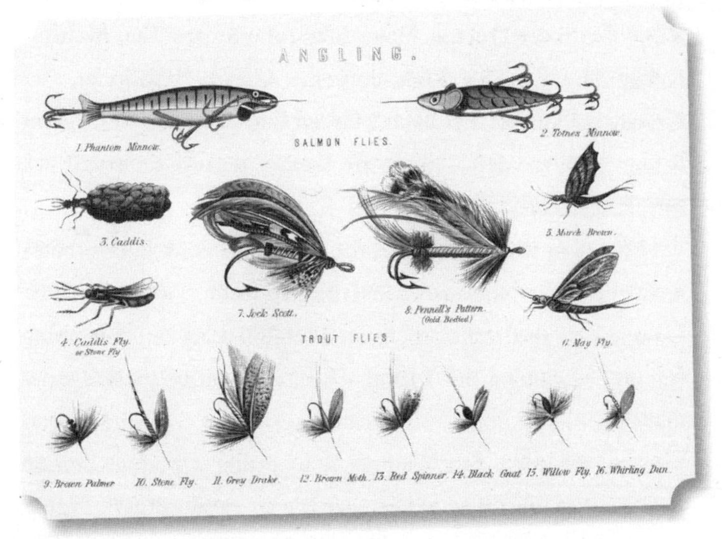

Paul: An early collection of salmon and trout flies with a couple of spinning mounts from less enlightened times.

it, they pickled most of the natural world and put it in jars). As the publishing industry grew, more people had access to books and could read up on the latest developments in fishing methods. And the study of entomology was as popular as it would ever be, with insects celebrated in art, fashion and design for their beauty and strangeness.

All these different things led to huge numbers of fishermen and academics becoming utterly obsessed with understanding every stage of the life cycle of particular insects, which they'd then try to replicate using whatever came to hand from feathers and beads to silk, furs and tinsel.

The flies came with fantastic names, all of which remind me of Bob's favourite cat names – the Bronze Pirate, the Fairy King, the Silver Doctor, Hairy Mary, the Stoat's Tail, Munro's Killer, Thunder and Lightning, the Green Highlander, the Grizzly Spider and McCaskie's Green Cat. I could go on and on listing these for days – there are literally tens of thousands of different varieties.

I've made my own flies in the past and I completely understand just how gloriously satisfying it can be. Tying your own variation on a pattern that's existed for 200 years and then going out and catching a fish with it – it's a little bit magic. It's especially pleasing if you can use materials that are just lying around – some thread, a few feathers – to create a perfect illusion that catches your dinner for you. It's an extraordinary experience from start to finish, and I can see why people become obsessed with it.

Apart from anything else, if you can get obsessed with the process of making flies enough, then you might find yourself in the lucky position where you don't then feel the need to go out and do the horrible, dirty old fishing bit. You could just sit at home, in the warm, go, 'Right, that fly's perfect, I'll put that in me box, put the telly on and reach for me drink.'

In 1895, George M Kelson's *The Salmon Fly: How to Dress It and How to Use It* convinced a new generation of fishermen that the best artificial flies needed to be made from exotic bird feathers. Partly, this reflected the fact that Britain had so many colonies across the globe, so these fancy feathers were readily available to those who could afford to buy them. Such was the demand for exotic bird plumage that when the RMS *Titanic* sank in 1912, the most valuable and highly insured contents in its hold was 40 crates of feathers.

To make a Jock Scott fly (named after its creator, a Scottish gillie), which is regarded as being the most remarkable, beautiful, intricate and hard-to-source of all the flies, and takes five or six hours to make, you needed feathers from the golden pheasant, guinea fowl, a peacock, a jungle cock and a fucking toucan.

Today, a newly made Jock Scott with genuine feathers will cost you upwards of £500. The original Victorian examples fetch huge prices today – not for use in fishing, but as works of art in their own right.

For most of us, if you wanted to make one of these Victorian beauties today, then you'd better hope the Natural History Museum have a tackle shop hidden away at the back.

Or you could do what Edwin Rist did.

In 2009, Rist – a 20-year-old flautist from Willesden Green who was studying at the Royal Academy of Music – broke into the Zoological Museum in Tring and stole nearly a million pounds' worth of bird feathers from their collection.

On a cold and dark Bonfire Night in 2008, Rist was escorted down to the museum's archive, where a quarter of a million bird skins were laid in storage. He had told the museum attendant that he was taking photographs on behalf of a graduate student who specialised in birds of paradise. What he was actually doing was casing the joint.

Six months later, Rist returned to the museum in Tring – only this time at night, scaling a wall, cutting through barbed wire, breaking a window, sneaking into the archive and frantically stuffing a suitcase with as many rare bird skins as he could. By the time he got back to Tring station at 3:00am, his case was bulging with nearly 300 rare and critically endangered Victorian bird skins, worth over half a million pounds.

Rist's crime – which at first seems utterly unfathomable – was motivated solely by his passion for salmon fly-tying. Since he was ten, he'd been making his own flies, but he'd been spellbound at a fishing show by a display of Victorian flies and dreamed of making his own. He'd become involved – this isn't a joke – with a shadowy underground online community of Victorian rare-feather fly-tiers, one of whom lived by the motto 'God, Family, Feathers'. Unable to afford these rare feathers to make the elaborate classic flies – a blue chatterer alone can cost over a grand – Rist decided to steal them.

What's even stranger about the story is that Edwin didn't fish. By all accounts, he'd never been fishing: he was just obsessed with the flies.

A year later, Rist was caught, and was sentenced to a one-year suspended sentence. Nearly half of the bird skins were never recovered. Rist added that he sold some in order to prop up his parents' failing Labradoodle breeding business – in less than a year, he made over £125,000. He's now changed his name and plays in an orchestra in Germany. It's the oldest story in the book.

A much more celebrated 20th-century fly-tier was Megan Boyd, an elderly woman who lived and worked in a rural village on the east coast of the Scottish Highlands. From her teenage years onwards, she spent her life making Victorian-style flies, supplying first the local fishermen and then, through word of mouth, outsiders. Ultimately, she ended up making flies for Prince Charles, who visited her at her little cottage on numerous occasions, even though it had no electricity or running water.

Boyd kept the Victorian fly-tying tradition alive well into the 21st century, but she was never an angler – she couldn't bring herself to kill a living creature, and, as a friend told a newspaper, 'She was there to catch the fishermen, not the fish.' Her own design of fly – called the 'Megan Boyd' – is famous for attracting salmon in summer when the water is low.

Boyd was awarded an MBE for her work tying flies (although she didn't go to the palace as she said she had no one to look after her dog) and after her death, her life became the subject of the

2014 feature-length documentary, *Kiss the Water*. Today, there are boxes of Boyd's flies in museums across the world and originals sell for thousands.

What's odd is that the majority of the most valuable flies that the Victorians created would never be used by any anglers today – not in a million years. And it's not just because they're so valuable and fragile that you wouldn't want to take the risk of losing one in the branches of a tree on the opposite bank.

They might be spectacular to look at, but they're entirely unsuitable for catching fish. In fact, these large, layered feathery clumps of brightly coloured feathers would be more likely to startle, spook and scatter any fish as a result of their totally alien appearance. You might as well chuck a lit Catherine wheel into a river (don't get any ideas, Bob).

For most of its history, the main form of trout fishing was what we now call wet fly – meaning the lure fished under the surface of the water, much like bait. But anglers on rivers with a chalk bed (like the Test, due to the clearness of the water) began to experiment with fishing the adult stage of the fly on the surface. Step forward, dry-fly fishing.*

Fishing with a dry fly saw the fly cast out to land on the surface of the water – it's termed 'dry fly' because it doesn't enter the river, as opposed to the wet fly, which (and you're probably ahead of me here) does.

---

* Other rivers like the Derbyshire Wye and possibly Driffield Beck in Yorkshire claim to have been the mother and father of dry-fly fishing. I'm not getting involved, okay?!

The technique was first proposed in print by WC Stewart in his 1857 book, *The Practical Angler; Or, The Art Of Trout Fishing More Particularly Applied To Clear Water.*

Stewart was a lifelong angler and in the opening of his book, he rages against the then-current perception of fishing by the general public:

> …there are few amusements which the uninitiated look upon as so utterly stupid; and an angler seems generally regarded as at best a simpleton, whose only merit, if he succeeds, is that of unlimited patience.

I think this feeling of being looked down upon is quite important to the history of fishing, as it influenced the way Stewart determined how to present his new techniques to the world: as a science, worthy of admiration and respect, and an art, practised only by those who were well educated in its complexities. Fed up of people thinking they were morons, Stewart did a remarkable 180-degree turn to reframe anglers as deep-thinking men of learning, using methods so technical that they were beyond the understanding of your common or garden non-anglers (bang on, Stewie, with one notable exception!).

Honestly, I think there's still some vestige of Stewart's defensiveness which hangs over the sport today – a lot of non-fisherfolk have an outdated sense that anglers are a tight-knit and unwelcoming community who are loath to bring inexperienced outsiders in. In my experience, the opposite could not be more true (save for a few fishermen I've met, who I'll admit were absolutely *awful* people. But that's not specific to anglers, it's true of humanity in general. You're just as likely or unlikely to meet a loathsome baker as you are a horrible fisherman).

Mind you, it doesn't help that anglers, like lawyers and criminals, have an entire vocabulary that must seem strange and alien to your average layman, nearly all of it introduced by the Victorians and only ever used in fishing circles.

We talk non-stop about trotting, rises, gentles, creels, fry and tackle. Some of the words we use don't even mean what you'd think they do – groundbait, for example, doesn't go on the ground, it goes in the river. Fly-tying isn't tying a fly onto your line, it's the process of making the fly from scratch. And that's before we've even got started on our bloody knots and flies. Carp fishing has its own language practically.

Fishing can seem utterly perplexing to the outsider, and you could make a case that the moment this intimidating language began can be traced back to that moment in 1857 when an indignant Stewart first put pen to paper.

What Stewart did was create a branch of angling which was different from the poor working man's humble fishing, and could thus be embraced by the aristocracy. This wasn't grubbing around in the dirt for maggots and having a fish force you to sit still for hours until it deigned to eat. Now fishing was about book learning! The thrill of the chase! The hunting instinct! Accuracy and sportsmanship! Deeply understanding the natural world! And not being poor!

In the late 1800s, Stewart's technique was advanced by the previously mentioned Frederic (FH) Halford, who wrote a number of hugely influential and acclaimed books on his new method. His work was once held in such high regard that Halford is known today as 'the Father of Modern Dry-Fly Fishing'.

In 1866, his *Floating Flies and How to Dress Them* became a huge bestseller – but his next book, *Dry Fly Fishing in Theory and Practice* (1889), became an aquatic bible to an entire generation of fishermen. Fishing the clear waters of chalk-bottomed rivers like the Itchen, the Kennet and the Test, he came to the conclusion that traditional wet flies looked nothing like the sort of insects that fish would encounter underwater, so he believed they were ineffective at best. Instead, he developed the upstream dry-fly technique.

While it most likely wasn't a totally new method, Halford's technique was to accurately mimic a fly landing on the surface of the water upstream of a trout. Trout face upstream, waiting for food to come down to them, so the fly would gently drift down. It would look exactly like the fish's regular source of food, being carried along by the flow. This, said Halford, meant they were more likely to rise.

His method also required incredible skill when casting, as you needed to silently land the fly at a spot directly above a rising, fish and hope it went along with the charade.

Halford believed casting a dry fly over a rising fish wasn't just a more effective method of fishing: he believed it was the *only* scientifically correct form of fishing.

> On one point all must agree, viz., that fishing upstream with fine gut and small floating flies, where every movement of the fish, its rise at any passing natural, and the turn and rise at the artificial, are plainly visible, is far more exciting, and requires in many respects more skill, than *the fishing of the water* as practised by the wet-fly fisherman.

Halford was utterly fundamentalist about his dry-fly method. He notes in his book that he would 'not under any circumstances cast except over rising fish, and prefer to remain idle the entire day rather than attempt to persuade the wary inhabitants of the stream to rise at an artificial fly'.

The dogmatic approach that Halford took in his books spawned a devoted cult of hard-line dry-fly acolytes, who followed his teachings with an almost religious fervour (it's no surprise that he was also called the 'High Priest of Dry Fly'). In short, if you were wet-fly fishing after his book was published, you were a heretic.

And, as with any major religion, even within his band of followers, there were schisms between factions: in this case, the purists and what Halford called 'the ultra purists'.

Those of us who will not in any circumstances cast except over rising fish are sometimes called ultra purists and those who will occasionally try to tempt a fish in position but not actually rising are termed purists (and I would urge every dry-fly fisher to follow the example of these purists and ultra purists). But in the early years of the 20th century, one brave soul stepped forward

Paul: Bob with a float/ maggot-caught trout. Halford and Skues will be practically fracking, they're turning in their graves so much. The big pike was caught on the fly on the same trip.

to challenge Halford's teachings. George Edward Mackenzie Skues (most commonly known as GEM Skues, which reads like the name of a really contemporary rapper out of Atlanta) was a lawyer and (some claim) the greatest fly fisher of all time.

In 1887, Skues bought a copy of Halford's *Floating Flies and How to Dress Them* and began experimenting with his own dry-fly techniques. As he went along, the former lawyer wrote about his new findings in the fishing press, rigorously examining every technique he used again and again until he was certain of his verdict (no doubt like Mr Mortimer did in his formative years in the law).

Skues collected his new articles in *Minor Tactics of the Chalk Stream* (1910) and *The Way of a Trout with the Fly* (1920), which caused a sensation in the world of dry-fly fishing. Firstly, he disagreed with Halford over sunken flies, which he believed were a legitimate tool in the angler's arsenal. But most controversial of all was his new creation: the nymph.

Skues invented a fly that imitated the nymphs – the larval stage of aquatic insects – that trout feed on underwater, which make up about 80% of their entire diet. His masterstroke meant he could silently slip his replicant into the middle of a big hatch and the ensuing feeding frenzy at the very moment the fish's guard was completely down.

If anything, Skues' method takes even more skill than Halford's. Upstream dry-fly fishing requires you to see a rise, cast to it, and then your fish might take it; with nymph fishing, you have to see right into the water, catch the little flash of the trout's open mouth, then judge where your cast needs to land to reach the fish below the surface and then decide if the fish has taken it.

Surprisingly, fishing with a nymph became immediately controversial, because it was seen as being too effective. It seemingly made the catching of trout a fait accompli. It was even called 'unethical' to use it, and the argument between Halford's disciples and Skues' gang quickly became fishing's version of Mayweather–McGregor.

One thing has remained constant since the days of Halford and Skues: whenever there's innovation in fishing, you can be sure that a big flare-up between the traditionalists and early adopters will follow right on its heels.

In 2008, trout fishermen using traditional methods became incandescently angry over the use of 'blobs'. Also known as attractors, blobs are brightly coloured lures that look nothing like insects, but when they're whipped through the water, trout cannot stop themselves lunging at them and chasing them until they're caught (allegedly).

But the use of blobs caused a storm that rages to this day. The England fly-fishing champion Chris Ogborne was quoted by the *Telegraph*, saying, 'Fly fishing is about imitating things that fish eat. Blobs are fundamentally bad for the sport. It's a very easy way of catching a lot of fish and takes the skill away. Any idiot can use them.' He was so furious about people using blobs that he resigned from the England team.

But the reaction comes from the same place as those who railed against the nymph: a love of fly fishing, and a genuine desire that future generations get to enjoy the same sport that's brought so many of us such long-lasting and total pleasure.

So if the history of fly fishing has taught us anything it's this: fish in a way that makes you happy. If you fancy trying fly fishing, give it a go. I hope you end up loving it as much as I do. If you buy a fly rod, the Fly Fishing Association of Great Britain aren't going to break into your house at night to make sure you aren't going out coarse fishing any more. No one's going to push you in the river if you can't cast perfectly on your first go.

Use a fly. Use a fly you make yourself. Use a nymph. Use a blob. Look, if you want to use a three-grand original Victorian fly made out of bloody dodo feathers, no one's going to stop you (unless it turns out to be one of those ones from Tring that the police didn't manage to recover. I can do you a few at a very reasonable price … shhhhh!).

THE BASICS OF FLY FISHING

Fly fishing is an extremely specialised way of fishing so all I'm going to do here is give you some of the basics.

There are a lot of books out there, which take you through the technical side of fly fishing in great detail, but until you've had a few hours at the riverbank, they're not going to open any doors for you. Knowing the theories of fly fishing inside out is all very well, but fish couldn't care less what you know about theory.

The only way you can learn how to fly fish is to fish with a fly fisherman. There's so much in the skill of casting a fly that can only be learned with experience. I can tell you what you need to cast, but it's the same as me telling you where to put your feet and then expecting you to be the prima ballerina for the Bolshoi Ballet.

There are some great online tutorials that can help a lot. In fact, I should practise a lot more as my technique needs brushing up and leaves a lot to be desired. You are always learning – or you should be!

One of the significant differences between bait and fly fishing is in the casting technique. The main one is that in bait or spin fishing, the weight of the float, ledger or spinner at the end of the line is what loads the rod when you cast. With fly fishing, the line *is* the weight. It's solely the weight of the line that loads the rod when you cast.

As such, fly-fishing line is much thicker and heavier than the line used in coarse fishing (there you use a very thin line – literally, the line of least resistance). So the line is what you use to cast your fly.

God knows what they're actually made of, but there are countless varieties. You get a low-stretch core with a plastic coating, you can have a floating line, an intermediate, a hover; you can have a very slow sink, a fast sink, an extra-fast sink, a tungsten core that will get you right down – there's the lot.

For a beginner, I'd stick with a floating line. Later, you might want a slow sink line for some lure fishing, but to start with, especially in rivers for trout, use a floating line.

There are various profiles of fly lines as well – they have a weight forward that loads rod a bit more easily, so you'd probably start with something like that.

When starting out, you want a rod of about 8 foot 6 inches to 9 feet. All the line and rods are rated. It's called an AFTM rating. That way, you can match the line to the rod. To start with, you'd want a line weight of about 6, so you'd buy a 6-weight line and a 6-weight rod.

The fly reel is a form of centre pin, but it's got a check on it. It's not a free running centre pin, it's always on a check, so you're not letting the line free spool.

On the end of your fly line, you attach a leader. This can be monofilament, fluorocarbon – a nylon of some kind – which starts out thick and then tapers away at the tip. You'd have at least nine feet of that, if not longer. Real experts can have a leader of up to 20 feet and might have one, two or three flies (known as a team of flies) on it, depending on the type of fishing they're doing and the rules of the fishery. Fluorocarbon sinks fairly quickly, so if you want your line to get under the water, you'd use fluorocarbon. But if you're dry-fly fishing you'd use monofilament, which sinks less quickly. To the other end of your leader you attach your fly or nymph. That's it in its basic form and absolutely fine for now. There is a bewildering array of leader types, materials and configurations that we can't even begin to contemplate here – in fact I'm feeling dizzy already.

You attach your leader to your fly line with a loop-to-loop connection. Fly lines these days almost invariably come with a factory attached loop, so you can tie a loop in the thick end of your leader; there's a knot called the perfection loop, which I'd recommend. Then you tie your fly to the thin end of the leader.

The last 18 inches or so are known as the tippet. Please use a grinner knot to begin with.

Starting out, you'd tend to just use one fly. Maybe a nymph (an artificial version of a subaquatic insect that lives its early life in the river, often replicating the moment it's ascending) or a wet fly (an imitation of a subaquatic version of a fly or a small fish) or a dry fly. Those are the basics.

Nymph and wet-fly fishing are where your fly breaks the surface of the water and floats down to a nearby fish. Dry fly is fishing the adult fly on the surface. It's very visual – you see the fish come up and take it.

The key is to present the fly precisely where the fish is rising, so that it goes for it almost as a reflex action. Some people say you should wait until you see a fish rising to a natural fly, imitate it and cast to it – these are the ultra purists that follow the works of FH Halford. Some fly fishers you meet will still swear by cane rods, as used in Halford's day, but if you ask me, the fish don't particularly seem to care what material your rod is made out of.

Lure fishing for trout is not to be ignored – in certain parts of the world, it's the main method used. It usually involves fishing sub-surface – so you might want a sinking line for that, or a long leader with a weighted lure.

Sometimes at the end of the season, certainly on some lakes and reservoirs, a highly exciting way to fish for big trout is with a floating fry. That's a very visual way of fishing as well.

The visual demands of fly fishing also mean you'll need polarised sunglasses, so you can still see what's going on in the water when the sun is dazzling off the river. Not only that, they are essential for safety.

Salmon fishing is its own world. You don't just get entire books on the basics of salmon fly fishing, you get entire libraries, and none of them will be of any use to you until you've learned the basic techniques of fishing as a whole.

Even when you've reached the point where you're fly fishing, prepare to be frequently bested by the fish. These are wild creatures who fight for their lives, and more often than not will end up slipping off your hook in a heartbeat. There's an unteachable skill you need to be a good fly fisherman and that's the mentality to lose a fish and get straight back in the game with as much confidence and spirit as you had before.

Fly fishing might seem complex, but like all the best hobbies, it is. There's nothing immediate about its joys – they reveal themselves slowly, giving to you, almost imperceptibly, a reward for the obsessive hours you devote to it. You'll never be the master of the art of fly fishing – but sometimes, on those rare special days, you'll fly fish like a master. That's what makes it so wonderful.

B: The tendency for Paul is to go fly fishing, so I tag along on occasion to go fly fishing with him. But when it comes to casting, I've still not cracked it. One in 15 of my casts is of any worth. But I look at Paul and the fishermen around me and it seems entirely effortless. He can land it on a sixpence.

P: It's just practice. Look at me, I've been doing it years, but I'm still not the best. You need to practise and I don't. In fact, on many days I think I've not learned very much. I often catch more trees behind me and bankside vegetation than I catch fish.

B: I think it's like the riding-a-bike thing. I will at some point do a cast and then I will be able to cast – but I haven't done that cast yet. One day, I hope I'll do that cast, and everyone around will shout, 'He's got it! Bob's got it!' And I'm really looking forward to that day.

CHAPTER 13

MORTIMER & WHITEHOUSE: GONE TO THE PUB

Bob: Are you going to the toilet in this one, Paul?

'There is nothing as yet been contrived by men, by which so much happiness is produced, as by a good tavern or inn.'

Dr Samuel Johnson, quoted by James Boswell,
Life of Samuel Johnson *(1791)*

Paul: Fishing and the pub. We like both and we like to combine them, don't we?

Bob: It feels like a very different experience to me, the pub to fishing. When you're catching fish, there's not the responsibility, I think, to be funny or interesting. But I always feel when you're in the pub, it's your duty to be interesting or entertaining.

P: Well, the most important thing is to get them in the right order. You need to go fishing first and then go to the pub. Because if you go to the pub first and then fishing …

BOB

Riverbank to pub – oh, it's an unbeatable combo.

We tease each other, Paul and I, when we're out fishing. It gets to four, five o'clock, which is a great time for fishing (especially for trout) but by that time, we're both just thinking of beer and pie. But we'll turn to one another and pretend we're still really into the fishing: 'No, no, we'll do another half hour.' We're desperate to go for the pie. 'No, come on, Paul, we should stay until we're sure there's nothing biting. I'm sure it won't be long now. Not long now at all ...' It's a process of self-denial that makes that pie even better. So by the time you head off the river for the pub, you really feel like you've earned it.

Paul recently thanked me for reintroducing him to the pub. He doesn't go so much any more – life gets in the way – but it's reminded him how much he loves them. I've adored pints all my life, but the pint you have after fishing is an entirely different pint to any I've experienced before. There are different pints, aren't there?

There's the beautiful one when you know you're going out on the piss, and that first one is just a cracker, isn't it? You're just up and excited, and that one is starting the ball rolling. That's a lovely one.

Conversely, the sixth pint when you're on the piss, that one can be a bit challenging and dreary.

Then there's the daytime pint where immediately after the first sip, you just wish you hadn't bothered. Completely ruins the day. You immediately want to pour it on the floor.

But there is a very specific pint I've discovered, which is the pint you have after you've been on the river for nine hours. And it's a spectacular pint. You only have it after you've been fishing. I call it the Fishing Activity Pint.

There are probably activity pints for other activities as well, but I don't do any other activities apart from fishing. If you were a squash player, I suppose that would be an activity pint too, but it would be very different from the river one. It would be refreshing, of course, but it would probably be in the municipal squash court, so that's not like the Fishing Activity Pint, because you'd most likely be enjoying it in a little country pub. The Fishing Activity Pint is all about the pub you go to afterwards.

I suppose hunting people and ramblers would also recognise that pub from their own experience, so maybe it's not the Fishing Activity Pint, maybe it's the Country Pursuits Pint. I think you could also experience this pint if you hired a boat at the weekend and went up the Thames, so long as you set the course for a pub.

We like to find a pub that's empty. That means you're going to be left to your own conversation and there's not going to be any trouble. There's nothing worse than going into a pub with one bored, drunk local lout in it, who's been waiting all day for someone to come in to hear his views on immigration and the beefs he has with other local louts. Or a rugby club who are six pints down and looking to have some fun with some old men from out of town. So we do like empty.

Although I say that, Paul and I, being Paul and I, we do also quite like it if the pub is *full* of real twats. So we can talk about them to each other and snigger. 'Look at that bloke, Paul!' We do like to sit there, whispering to each other and being rude.

We go to these remote little pubs, particularly in Norfolk and Wales, and there'd usually be an old geezer who'd come over and say hello. They're from a different era, so they don't want a photo with you, or for you to say hello to their girlfriend on the phone – but they might have seen the show and they want to give us some real local knowledge about the fishing there.

There's a little community of people you meet in the pubs when you're out fishing. They'll come over, give you some advice, have a look at your gear, tell you their story. 'All right, Paul, I've seen you on the telly. Now, have you fished here before? What's your bait?' It's a nice way to interact, over a hobby. It's as old as the hills, but it makes the conversation so easy.

When we were on the Herefordshire Wye, fishing for barbel, we got talking to a bloke in the bar. I think his name was Banjo. He told us, 'You want to use some black pudding for bait. It's been doing really well.' So we got some black pudding the next day and tried it; I lobbed some in. Didn't catch a thing on it. But I enjoyed that. It was a little exciting narrative that made the whole trip feel like more of an adventure. 'I met a bloke called Banjo, he told me to use black pudding, and I caught a 4lb barbel.' In the end, that's not how it panned out, but you can't have it all.*

---

* Paul: Meanwhile, Banjo is probably, telling his mates: 'Them two idiots, Mortimer and Whitehouse, was in last night. Told 'em to use black pudding as bait and they fell for it! Wankers!'

The next important thing is the pub's got to have really nice beer. If you're experienced with pubs, you can just look through the window and sniff out what they're about. Whether it's a sports one, it mainly does food, or a hangout for layabouts: if you look through that window and the beer pumps are all illuminated, then that spells lager and chances are there's no real ale selection. If you see a real fire going, that's a winner. Same if you can see there's a big shaggy old dog asleep on a huge flagstone. Or a chalkboard with a load of local beer names written on it, and some have already been rubbed out. Get straight in there, it's going to be perfect.

Essentially, it's pubs straight out of *An American Werewolf in London*, but with no one in.

My little ritual would be to find a lovely little pub in Norfolk or Somerset or Wales – oooh, I can see them all now. Standard, Paul and I will get there around half-seven, and I'll start with a pint of the best.

I'm still okay with beer – even after my heart op, I was never told not to drink. When I first sat down with the doctors, they gave me the standard spiel: 'Now, don't smoke, don't drink, don't do anything bad.' But when I saw them later on and talked things over with them, it turned out the advice not to drink is more part of their general medical advice which they slip into everything, and it's not specific to having heart disease. Paul's never been told not to drink by any of his medical specialists either. They do give it as general advice, and they're probably right too, but it's not heart specific.

So I'll start with a pint of the best bitter. I reckon I like to do three pints. I'll have a pint of best and then move on to have a couple of pints of their local craft ales, if they've got them.

I really enjoy a craft ale. I do a bit of beer brewing with a fella, and the soft, watery English hop beers aren't as prevalent these days. They don't seem to be what people want. So we're using a lot of American hops, which give you a more fruity, citrusy, grapefruit taste and a slightly more bitter beer. I worry for the future of the delicious traditional pint of bitter. The brewer I work with does an IPA and a Pilsner, but he doesn't do a best bitter – a classic 3%. It would be awful if it disappeared.

Then I'll buy a pie.

Paul's a serious wine man. Serious to the point that when we go to a pub, he'll ask to see a wine list. That's Paul's big indulgence, nice wine. I *hate* wine. With beer, you can have five pints and make a nice relaxing night of it. You can't do that with wine. Five wines and you'll be absolutely gone. In fact, two glasses of wine is all I need to start getting hot and bothered.

The best pub day we ever had was after we'd filmed the episode about the barbel. They're fun days, but they're long days, and Paul and I went into this boozer, ordered a pint of best each, sat down and neither of us spoke for 20 minutes. We couldn't have been happier. It wasn't contrived, we just noticed that neither of us had said anything. That was a completely magical moment and they don't come along very often.

When it comes to the hotels we stay in, I'm like everyone else. When I walk into a hotel room, I'm not looking for anything else but a decent telly. I see a nice big one, I'll say '*Yes!*' Then I'll walk through to the bathroom to see if it's got a bath, because I'm not a shower person. So it's got a bath – '*Yes!*' And then I might lie on the bed and see how irritating the groaning sound that comes from the air con is. And then that's fine by me. I don't need anything grand. When we're on tour, Jim and I, you always get asked if you want to stay in these fashionable country house hotels. But we always stayed in the Holiday Inn. They've got good tellies – I think with a bit of Sky as well – and a bath. I don't like the hotels where there's no bar. An honesty bar, that's sort of shit. I don't like people coming over when I'm having my breakfast to ask if I'm enjoying it.

If you asked me what the perfect place to stay would be, Paul and I would stay at the Greyhound in Stockbridge. It's basically a

Bob: Paul and I enjoy one of the lovely hotels we had the pleasure of staying in.

few pleasant rooms above a really great pub, with a lovely restauranty bit and a bar that's open to the high street and open to the community. A fishing hotel/pub par excellence. It helps that it has some fishing on the Test in its back garden! That's a great combo. We went to one in Abergavenny called The Angel, and that was similar – big old fire, big tables to eat your food on; oh, they'd got it sorted. We really liked that one.

I suppose in some ways it feels like a slightly smug retirement present for Paul and me. It feels like we're getting to go on holiday in an episode of one of those Richard E Grant shows where he goes to Britain's best hotels, or climbing into a *Sunday Times* pull-out. I'm sure you get much posher ones, like where Elton John would go, but we get to go to all these lovely pubs and these lovely hotels.

It's sort of like this is our reward for working all those years. And it makes it feel like it was all worthwhile.

P: Did you mention the sex pub?

B: That had a very sexy vibe.

P: We went pike fishing, so we got in touch with John Bailey. John knows the area so we asked him to recommend us a hotel. We said, 'Can you find us a nice place, with nice rooms, a bar and a restaurant?' And so we ended up in this place. It was a lovely drive up, very cold out, with a bit of snow on the ground.

B: And then we turned off into this car park area at a sort of motel.

P: It looked fine, it had a nice bar, the rooms were lovely…

B: Mine had a sauna in it.

P: That was the moment a little alarm bell went off: 'Hang on – why have I got a sauna in my room?'

B: I had a walk around and at the end of the corridor was a big door, with one of those yellow plastic signs outside it, like they put out when the floor gets wet. But it didn't have a slippery floor warning on it. It said, 'Don't come in – kissing and cuddling going on here'.

P: Kissing and cuddling going on!

B: So I looked through the smoked glass window on the door, and there was a hot tub which, I believe, was full of couples having it off.

P: Were they actually …?

B: I assume so. I didn't actually get in there to see.

P: We went down the next morning, and you know how in rural hotels at breakfast, there's always old couples? *Not. One.* Just me and him.

B: It must have been a sex place. To have such a homogenous clientele …

P: There were only youngish couples dotted around.

B: So it was just couples who were having sex and us two old dried apricots.

P: The two of us came down to breakfast, and waved to them all: 'Good morning! Morning! What are you up to today? Group sex? Right-o! No, that's very kind, but we're going pike fishing! Better put the old thermals on!'

CHAPTER 14
THE END

Paul: Come on, Bob, off into the sunset we go…

'Why do old men wake so early? Is it to have one longer day?'

Ernest Hemingway, The Old Man and the Sea *(1952)*

BOB

I had an awful thought when I was walking down to the train station the last time I was making my way over to see Paul. I thought, 'God, what if you lost your fishing buddy?' What if Paul snuffed it? It would be awful for me.

It's terribly sad when old fishermen part, isn't it? You see two old blokes fishing, in their sixties or seventies – at some point they're going to lose their buddy.

I don't know if I'd go fishing on my own, because although it's hackneyed, it's absolutely true: it is about companionship. Even if you're not speaking, it's important that your pal is there, just up the bank. 'Oh, I wonder how he's getting on' or 'I'll go and see how he's doing' and you meet up and go to the pub of an evening.

I might be wrong but I can't see that I'd go fishing if Paul wasn't with me. It would hammer home the solitude. I don't know what it would be about – you drive up on your own, go to the bank on your own … I think I could get some pleasure out of it, but if I did go fishing again, it would be for different reasons. If I went again, I'd only be doing it for old time's sake, remembering when I came with Paul.

We fished on the River Tay for salmon once, and there was an old guy there. We got talking and he told us he was 76 and he had cancer, and it had just blasted through him. And he was just waiting to die.

I asked him about it and he said, 'I really haven't got long. It could be six weeks, or it could be tomorrow.' So I just asked myself that basic question that everyone must ask when they get a diagnosis that's terminal: 'What's the best way I can use the time I have left?'

But he said, 'I haven't had the worry of wondering how I should best use this time. I was so pleased that this wonderful hobby has given me the answer.'

If I was asked that question, I wouldn't have a clue. 'I want to see my kids, I want to spend time with my wife, I want to go to that place I've never been to …' I don't know how you'd face that.

He made a hard, fast decision: he was fishing the Tay until he dropped. By the time you're 76, you know what makes you happy. So he'd gone to the Tay every day. He was probably thinking about lost mates he used to fish with.

And he did exactly what he said he would, and he died about three weeks afterwards. But with his time on earth running out, he knew exactly what he wanted to do.

He wanted to go fishing.

ACKNOWLEDGEMENTS

Paul: Bob and I would like to thank David Brindley, Daryl Ewer, John Hall, Patrick Holland, Jason Lewis, Rob Thompson, Nicky Waltham, Lisa Clark, Jamie Munro and all those involved at the BBC, Greenbird and Owl Power; Gordon Wise at Curtis Brown; Marc Haynes; Joel Simons and Karen Stretch at Blink Publishing; the brilliant John Bailey; and all the production team and crew of *Mortimer & Whitehouse: Gone Fishing*. Personally, I would like to thank Alexandra Cann and Jacquie Drewe; all at CP and R; Dr Bucknall; Mr Rob Hutchins; Mr Tan Arulampulam. My love to Carys, Molly, Lauren, Delilah and Mine.

Bob: And I'd also like to thank Keith Bridgewood, my oldest friend, with whom I shared all my childhood fishing memories; my medical heroes, Dr Bob Bowes, Dr Clive Lawson and Mr Chris Young – I know exactly where I would be without them; and my boss Caroline Chignell.

BIBLIOGRAPHY

Parker, E., Turrell, W.J., Ensom, E. & Others. *Fine Angling for Coarse Fish*, Seeley, Service & Co., 1930

Paxman, Jeremy (ed.). *Fish, Fishing and the Meaning of Life*, Michael Joseph, 1994

Ransome, Arthur. *Rod and Line*, Jonathan Cape, 1929

Venables, Bernard. *Mr Crabtree Goes Fishing: A Guide in Pictures to Fishing Round the Year*, Mirror Features, 1954 (fourth impression)

Walton, Izaak. *The Compleat Angler*, 1676 (fifth impression)

BONUS CHAPTER

ON THE RIVERBANK WITH BOB AND PAUL

'What do you think costs more from Boots, a pack of two fresh and breezy insoles for your training shoes, or a small 200ml bottle of infant Calpol?'

Bob Mortimer

Enjoy this bonus chapter in which Bob and Paul head to a fishery in Kent for an afternoon on the riverbank…

Bob: If people knew what the reality was fishing with us, they'd get a bit of shock. Should we show them? Fishing without cameras.

Paul: Fishing without cameras.

B: Fishing without cameras.

P: Well, then they won't be able to see us.

B: Go on, let's do it.

P: All right. You going to bring your blue rod?

B: Me blue rod will be in front of me, yeah. Me fly rod, behind. Come on.

P: All right, lovely. Come on, then. Isn't this exciting, Bob?

B: Where we going to go? Do you know somewhere?

P: I know a few places. I'm going to surprise you.

The distant rumble of a car engine as it drives down the towpath and into a car park. The engine stops, and the doors open…

P: Well, thank you, Bob, for excellent driving.

B: Well, here we are. It's lovely, isn't it?

P: Breathe the fresh air.

B: Lovely. Makes a change.

P: But we're only hour, hour and a half out of London.

B: Yeah, and everything changes.

P: Doesn't it?

B: Yeah.

P: We can crunch across the gravel. And then we will stealthily approach the water.

B: Let's paint a picture for the readers: we're in a valley.

P: We are. We're in a sylvan valley. Do you know what that means?

B: Nope.

P: Wooded.

B: We're in a wooded valley. A sylvan valley.

P: Sylvan, isn't that a nice word? You could use it in one of your poems, couldn't you?

B: I can see what appears to be a river.

P: Yeah there's a little river drifting off down there. I'm not even sure what river it is; I did know what it was, but I've forgotten it.

B: You've forgotten it? That's just your age, Paul. Well, there's three other people fishing here, they all look like they know what they're doing.

P: They don't really know, I can tell you that now. Anyway, I've seen a rise up there, or is that ducks? I think it's ducks.

B: Is there just rainbow trout in here, Paul?

P: It's mainly rainbows, but guess what trout are in here? Spartic trout, how about that?

B: What on earth is that?

P: I am Spartacus!

B: It's trout with attitude!

P: Yeah, really serious attitude; apparently, it's some kind of cross between a char and a rainbow trout.

B: And a char itself is a cross isn't it?

P: No, Arctic char is a pure species in its own right. Anyway, we're going to fish with a floating line, a longish lead, not that long a leader and we'll be using a weighted nymph. Okay? Have you actually ever used one?

B: Yeah, I've used it in the TV series.

P: Now I just need one rod really, it will just complicate it if I have two. So shall I go for a light line approach, Bob? A lighter approach than you, son?

B: But why would you go lighter, I've never understood the lighter to—

P: You want the more heavy, you want me to go hardcore? Actually there's quite big fish in here yeah let's—

B: Would you refer to me when I'm not here as your 'fishing mate', your fishing 'pain in the arse', or your 'fishing nuisance'?

P: 'Fishing nuisance' is good. Yeah I like that, I'll go with that. 'Fishing halfwit' is my usual thing that I go to. It's not very nice, is it, it's not fair on halfwits.

Silence. The two men continue to prepare the line.

B: Well, thank you for setting this up. You're putting a leader on the end of the fly line at the moment?

P: Well, you've got a leader on. I'm putting a tippet on the end of your leader.

B: Oh, and you're going to do your fancy knot, aren't you?

P: I'm going to do me grinner. It's the same knot I'm going to use to tie your fly on as well. Are you cold?

B: Yeah, but I've got a hat on.

P: Have you got one of your comedy hats on?

B: Well, it weren't meant to be comedy, Paul. I often thought I cut a dash in them.

P: Well, you just about get away with that one.

Silence. The sound of quiet reflection as the two men continue to prepare the lines.

P: Right, this is a three-turn water knot.

B: Okay. Yeah.

P: I'm just going to put that through—

B: And that's joining one bit of line to another bit of line?

P: Yeah. But while talking to you I'm not sure that I did it [checks the line] Yeah, it's all right. Bob you're going to use a weighted nymph, all right?

B: Will it go to the bottom of the lake?

P: It probably won't because the breeze will pull your floating line round; but that's okay, we'll just let it drift round on the breeze. You can try one of these [the nymph], do you want to try this? Does that appeal to you?

B: It's black and gold.

P: Yeah, black and gold, who can resist that combination, black and gold?

B: Throughout history it's been—

P: If I was a knight and I wanted to strike fear into the hearts of my enemies I would go for black and gold, that combination, wouldn't you?

B: The Black Knight, black and white, black and gold.

P: Yeah, so you want to try this, you want to give this a go?

B: Yeah because I think that you know that it's good 'un. But I'm not putting it on you saying, 'Oh well it's your fault' if it doesn't work.

P: Look, it'd be daft for us both use exactly the same method, wouldn't it?

B: Oh you're going to use something spectacular, aren't you?

P: No, I'm not. I'm giving you a black gold head nymph and then I'll use maybe something a bit drabber or a bit more like a hare's ear. I'll show you what that is in a minute.

Paul casts out on to the lake. It's a beautiful sunny spring morning and the zip of the line pierces the air.

B: So we're on quite a wide grassy bank here, Paul?

P: Yeah.

B: Can the fish see us from where we are stood?

P: They probably can, although the fish that we've seen are quite a long way out, I don't think they'd be particularly bothered by us out there in the main bowl. But obviously the fish that come into the margins would see us. So if we were going to go and fish out there around the edges I would say stay back from the edge and crouch down. I mean sometimes the fish will follow your lure right in. What do you think they're thinking?

B: Well, their minds are just full of greed for food, isn't it, I imagine.

P: Is it greed or curiosity? Would you follow a hotdog out of curiosity as well as hunger? Say in your conventional household, Lisa [Bob's wife] put down a hotdog and then walked out the room and there's the hotdog there and it started to move. Would you go for it?

B: Do you know, I'd be tempted to.

P: You might, mightn't you; you'd grab it, wouldn't you? Similarly, a fish might think, 'Oh, what's that?' And if it's getting to the edge, he thinks, 'I'll grab it before it gets away.'

B: I think my grabbing would be partially based on ownership, like that's *mine*. Where do you think you're going? You belong to me.

P: Yeah, that's well put, I like that. Yeah.

B: But I don't know if the trout feels a personal ownership.

P: It might if it thought there was competition for that same morsel. Say I came in, right, into the room, and you could see me going for that hotdog, you'd definitely eat it then, wouldn't you?

B: Oof, yeah.

P: Yeah, if you saw Whitehouse—

B: Oh, I'd destroy it, you're not having it.

P: Yeah. That sums up the human condition, doesn't it? You would destroy that hotdog rather than see me get it.

Silence. Bob watches intently as Paul continues to cast out into the lake.

B: [Bob points at the lake across the fishery] So this lake, across from here. Would you say it's about 150 feet wide?

P: I'd say it's more than that. I would say it's about 100 metres isn't it? Maybe 70?

B: Usain Bolt would run that in about four seconds.

Silence. Paul senses a bite.

P: There's another fish rising. Keep that rod down, don't move the rod. Yeah, that's it; like that. Just watch that line; you might see it shoot off or it'll tighten.

B: And then I would lift the rod, yeah?

P: Yeah. Don't need to strike hard, just move that way, okay.

B: You don't play it with the reel, do you, Paul? You play it with the line.

P: I try to get the line back on the reel, but often when you're fly fishing you have a lot of line out. It requires a lot of practice casting. It's analogous to the golfer's swing and you know how bad I am at that? You were quite good at your golf, Bob, weren't you?

B: Yeah, but I felt I had mixed feelings about it, Paul, because you all thought that I'd play golf before.

P: Yeah, I'm actually convinced the way you hit that ball that you've played golf before.

B: I haven't, Paul. No.

P: Honestly, I know that you've got good hand-ball-eye coordination because you used to be a pretty useful footballer … [Paul spots something in the water] There's a little buzzer! There below.

B: There he rolls.

P: We can target that fish in a minute.

P: Damn. I thought I had a chance there with that cast. And now I'm in!

Paul begins to bring the fish in

B: Well done, Paul.

P: It's a good fish.

B: Keep on top of it.

The fish is a fighter! It makes a dash for it!

B: Whoa, did you see him go? Whoa, he's feisty.

P: He is, he's airborne, isn't he? Airborne division.

Another break from the fish. This one's even bigger!

B: Whoa!

P: Or he's got in some weed there. I said that cast might work, didn't I?

B: Sometimes you just know, don't you?

P: Yeah, this is when it all lands properly and you know there's fish moving out there as well, like that. Oh, he's not coming in yet, is he? Bit of life in him, ain't he?

B: So if he wants to go on the run you let him, yeah?

P: Yeah, that's right. But he's not that big is he?

Bob lifts the fish out of the water with the net.

P: There you go. Well done, Bob. Wow. There you go.

B: So there he is.

P: Yeah.

B: A two-pound rainbow?

P: Yeah, I reckon don't you?

B: I reckon.

P: Feisty, isn't he? Look at that. It's beautiful.

B: Yeah, isn't he pretty? So Paul, with all that leaping from this one, will all the other fish have cleared off now thinking something's up?

P: Yeah, I mean it will disturb them. Sometimes they follow it in, they're not quite sure what's going on, you know. It's like, let's say we're in a nightclub together and I've said the wrong thing (which is quite possible, isn't it) to a couple of blokes and they're battering me – whack! – like that. And you're at the other end of the bar, there's a very attractive lady there with you, your beer's there, and you're thinking, 'Shall I go and help him or shall I just nothing to do with me and carry on, err, feeding?' What would you do? I know what you'd *pretend* you'd like to do is you'd like to wade in, deck the geezers and then go back to your beer and the attractive lady.

B: I think I'd inform security that there's a chap in trouble over there.

P: It's for the best.

B: That's for the best, isn't it?

P: And you carry on with your wooing?

B: Yeah.

Something catches Paul's eye.

P: There's a very fast duck coming at us.

B: This is the fastest duck in Surrey.

P: I've never seen a duck go so fast.

B: Lottery winner.

P: Yeah.

B: He's going down the Post Office to get his winnings.

Paul returns to his rod.

P: Do you want to try a buzzer or a dry fly?

B: A dry fly.

P: A dry fly. Yeah, you were asking about whether that fish will have been disturbed, but you can see that they're still rising out there, because I hooked that fish quite a long way out and we landed it in here, in this little bay; this is where it was doing most of its acrobatics, wasn't it? I think we've seen a couple of fish rising out there, so you want to try a dry fly now?

B: I do, yeah. I can sense from your attitude that you think I'm being a plonk.

P: Actually, I think these fish are feeding on buzzers just under the surface, so would you like to try that or do you want to try something else? You want to put a daddy long legs out there don't you?

B: Yes.

P: Come on, then. Right we're going to put a daddy long legs on just for you—

B: Put a clonker on it, yeah!

P: I'm going to put one on, mate, for you, all right?

The clonker is attached and the line is cast. They wait.

P: The thing about the daddy you're casting from here, now the wind's dropped you should be all right. Because it's quite a large profile, isn't it?

B: It's kind of just an enticing mess, isn't it, in a way?

P: It is a mess of legs and body, isn't it? But it's a nice mouthful for a fish because it gives a lot of protein in one bite.

B: Yeah.

A blackbird tweets in the background. Suddenly a bite…

P: Oh Bob, look, look. Yeah. Ahhhh. To the dry daddy as well. Oh my God!

B: Check out the daddy doing the business though!

P: I know.

But the fish comes loose.

B: Fuck!

P: Start again.

B: So talk me through that, Paul. We weren't quite concentrating?

P: What happens when we don't concentrate?

B: You lose a fish.

P: You lose a fish. There we were, we were talking about the daddy and I was about to lay into that blackbird.

B: You were, yes.

P: I really was. Whatever Paul McCartney thinks about them, I find them really annoying. Eeh, eeh, eeh, eeh, eeh, all night, early morning. And I was just about to go into my rant and I was handing the rod over to you and what happened?

B: Bang!

P: That fish took that daddy and we were—

B: I yanked it out like—

P: You yanked it out mid-handover. And that'll teach me to disrespect the countryside and all its inhabitants, won't it?

B: And it'll teach you to stop disrespecting the daddy, that's the go-to fly, Paul.

P: I know I'm slagging off the blackbird for its piercing, shrill one-note song, and I know it has other notes in its armory, but it's a joy really, isn't it, to be out of the hurly burly of the urban environment, Bob.

B: Of the urban environment, yeah.

P: And by a bit of water.

B: And with two hours by the bank you just begin to relax into it, don't you?

P: It's interesting, isn't it, because you're just sat here…

The quiet of the lake returns. The men inspect their quarry and then return to the rods.

B: Do you know your chemists? Do you think you know your chemist shops?

P: Do I think I know my chem … What do you mean the products or the—

B: The products, the pricing.

P: I sense that you do, so you're going to hit me with some really difficult questions now about chemists, aren't you?

B: What do you think more costs more from Boots, a pack of two fresh and breezy insoles for your training shoes, or a small 200ml bottle of infant Calpol?

P: Now you've put me on the spot. How disposable are these insoles, I'm calling them?

B: Basic Odor Eater insole, I don't know what more I can say. Charcoal-infused or something.

P: £3.99 for the Odor Eaters and £5.99 for the Calpol.

B: Paul you are very good. £3.99 for the insoles, correct. The Calpol's £6.99.

P: I nearly went for £6.99 you know. Wow.

B: All right. Do you know your chemists? This is a challenge: do you know your chemists, yes or no?

P: I didn't do bad there being put on the spot.

B: So I ask you again, do you know your chemists?

P: While studiously fishing a Diawl Bach?

B: Do you know your chemists? To activate the quiz you have to say, 'I know the chemists.'

P: I think I have to say 'yYs, Mr Mortimer, I know my chemists.'

B: Okay, which is more expensive a 12 pack of 200 milligram Nurofen?

P: A 12 pack of 200ml, yeah.

B: Or a replacement mop head?

P: Which chemist are we talking about here? Because I know chemists that sell Nurofen, but I don't know many that sell mop heads.

B: Actually I should have said do you know your chemists/ hardware shop? Do you know your B&Q? Actually you wouldn't be able to get Nurofen there, would you?

A slight nibble. The fish disappears.

P: Ahhh. Bloody hell.

B: Bloody Nurofen.

P: Fucking Nurofen.

Silence.

B: If you lived in America you'd have the look of what they would call a 'grifter', do you know what I mean?

P: What, like a hustler?

B: Well, like poor stock, mobile home, you know. Trailer park.

P: Living on my wits trying to—

B: Yeah, but you had a bit of brain about you.

P: All right, yeah, so I'd sneak a buck out of people, yeah. [Adopts American accent.] 'Hey could you lend me a dime, buster?' That sort of thing, 'Don't worry I'll give it straight back.' And then I'll do some sort of trick on you.

B: Yeah, in the Wild West you'd have sold glass eyes and stuff.

P: I don't think I'm very good at that sort of shit, actually, Bob.

B: I'd have been a doctor.

P: What sort of doctor?

B: The type of doctor that you go and see if you're ill. That one.

P: Oh that type of doctor, the conventional doctor. Ah. Right.

A lull. Paul jerks the rod.

B: Do you know your Marks and Spencers?

P: How come you're so good at all this stuff?

B: I'm going to give you the most basic Marks and Spencers one, right.

P: Go on then.

B: A pack of six black cotton socks, yeah.

P: Cotton rich?

B: That's exactly as they're described, yeah. A pack of six of them, or a pack of four basic cotton rich Marks and Spencers boxer shorts, I think they call them trunks – you know the underpants that have got a bit of leg?

P: I think I might be wearing some right now.

B: Okay, so price them two up.

P: Cor, you're a hard taskmaster, aren't you?

B: Thank you. I've got a reputation to keep up.

P: Okay. £7.99 on the socks.

B: Yeah.

P: And how many in the pants packet?

B: Four.

P: Okay. £12.99.

B: You're very, very good, Paul.

P: Am I?

B: Yeah, 12 quid for the pants. And eight quid for the socks. I mean, it's very good.

P: It's pretty impressive, isn't it?

B: Give me five really simple fishing questions. Let's see how much I've learned—

P: Oh, okay.

B: Really simple, you know—

Suddenly… A bite!

P: Whoa. Did you see that one take? Yeah look. Cor.

B: Wow.

P: Wee!

Paul begins to reel it in.

P: All right, question number one. What fly am I using to catch this lively rainbow?

B: A call belly, or whatever it is.

P: Whoa. A call belly? The Diawl Bach, yeah.

B: The Diawl Bach? That was an unfair question. I was talking basics.

P: All right, okay, what type of line … Whoa, look at this one go. Bloody hell, Jesus. It's like a sea trout, look at this. He's got some poke, hasn't he? Bloody hell.

B: Did he rise to take that, Paul?

P: No, he took under, he took it with a bang. Easy. Whoa. Bloody hell. He's in good nick, isn't he? Look at that he's a bar of silver, isn't he? Got him. Nice. Yeah he's a bit chunkier isn't he?

B: Yeah.

P: He's a good fish, Bob.

B: Oh, he is bigger, isn't he?

P: Yeah, he is bigger, yeah.

B: So that was taken on the bal—

P: On the Diawl Bach. Told you, I've had about three casts and two fish. And there we go.

B: Well done, Paul.

P: My heart rate probably went up a little few notches during that little period there, did yours?

The men inspect their catch. Then Bob casts and begins to fish.

B: So, Paul, you'd have to say two beautiful rainbow, three near misses.

P: Yeah, and a lot of excitement too, Bob.

Moments later Bob gets a bite.

Paul: And you're in, he's in. Yes. Let it go, let it go, let it go, let it go, let it go, let it go.

B: Don't you have to keep in touch with it?

P: Yeah, you do, but we're going to get that line on there, quick. Right, now you can wind in. Wind in, wind in, wind, wind, wind, wind, wind. Let it go, let it go, let it go, right now, let it go, okay.

The fish, still attached to Bob's line, swims as far away as it can.

B: Well, he's going to go into the next county if you're not careful.

P: Right let me stop him a bit. Wind, wind quick. Right he's going to run again in a minute.

B: He is n' all!

P: Yeah he will. And he's going to jump. Right, watch it now. There he goes – that's it! Let it go like that. Whoa.

B: Is he gone? He may have got off.

P: No, no, no he's still on. All right lift your rod up a bit, let him go, let him go. Whoa. [Paul reaches down with the net and fishes out the catch.] Bit of quick work with the net there. Look at that. Well done, Bob. Bloody hell.

B: My first rainbow!

P: Look at that, that's a clonker isn't it?

B: Isn't he? So just as I was saying two fish, two become three.

P: Two become three. It's like a Spice Girls song, isn't it? Cor, you've caught the biggest of the day, Bob. On the Diawl Bach. Do you know what you call two fish?

B: I don't know.

P: A brace.

B: A brace.

P: What would you call three fish?

B: A trio?

P: A leash.

B: A leash of fish? A leash of trout.

P: A leash, yeah, and we got a leash between us.

B: Well, what a lovely end to the day.

P: That's a beautiful full stop, isn't it?